CW01262120

EVERY LAST FISH

Also by Rose George

Nine Pints
Ninety Percent of Everything
The Big Necessity
A Life Removed

Every Last Fish

What Fish Do for Us and What We Do to Them

Rose George

GRANTA

Granta Publications, 12 Addison Avenue, London W11 4QR
First published in Great Britain by Granta Books, 2025

Copyright © 2025 by Rose George

Rose George has asserted her moral right under the Copyright,
Designs and Patents Act, 1988, to be identified as the author of this work.

All images by Rose George except p2 Wellcome Collection,
p44 Matt Harrop, p64 National Transportation Safety Board, p87 Wikimaribarre,
p124 Wellcome Collection, p140 Estate of Petty Officer Harold Dodgson BEM,
p172 Open Seas, p188 Environmental Justice Foundation,
p204 Christopher Hilton, p242 Kim Hansen.

All rights reserved. This book is copyright material and must not be copied, reproduced, transferred, distributed, leased, licensed or publicly performed or used in any way except as specifically permitted in writing by the publisher, as allowed under the terms and conditions under which it was purchased or as strictly permitted by applicable copyright law. Any unauthorized distribution or use of this text may be a direct infringement of the author's and publisher's rights, and those responsible may be liable in law accordingly. Please note that no part of this book may be used or reproduced in any manner for the purpose of training artificial intelligence technologies or systems.

A CIP catalogue record for this book is available from the British Library.

1 3 5 7 9 10 8 6 4 2

ISBN 978 1 78378 792 0 (hardback)
ISBN 978 1 78378 794 4 (ebook)

Typeset in ITC Berkeley Oldstyle by Iram Allam
Printed and bound by CPI Group (UK) Ltd, Croydon, CR0 4YY

The manufacturer's authorised representative in the EU
for product safety is Authorised Rep Compliance Ltd, 71
Lower Baggot Street, Dublin D02 P593, Ireland.
www.arccompliance.com

MIX
Paper | Supporting
responsible forestry
FSC® C013604

www.granta.com

For those in peril on and in the sea.

And also for Thomas Ridgway, the best of men.

Me: I don't eat fish.
Fisherman: That's a good start.

*Information technology could accelerate exploitation –
helping fishers track down every last fish.*

FROM A BLUE PAPER FOR THE HIGH LEVEL PANEL
FOR A SUSTAINABLE OCEAN ECONOMY[1]

Contents

1. Dora's Breakfast — 3
2. A Man in His Boat — 23
3. Fast Fish Food — 45
4. We're Rolling Over — 65
5. A Coarse and Vulgar Woman — 87
6. Trawling — 107
7. Fish Are Not Chips — 125
8. A Tin-Can Navy — 141
9. Before, There Was Fish — 155
10. Down Below — 173
11. A Dedicated Fish Warden — 189
12. A Very Slavery Job Actually — 205
13. A Nice Bit of Salmon — 225
14. Windy City — 243
15. Fish In, Fish Out — 255

Acknowledgements — 262
Which Fish Guide — 264
Notes — 269
Select Bibliography — 299
Index — 302

EVERY LAST FISH

Two fish watercolour, 1783

1
Dora's Breakfast

I am 10 metres underwater and I feel fine. I shouldn't feel comfortable: this is not my element. The ocean is no longer for me: it has been many million years since we evolved from fish. I am in the most alien of environments I will ever be in because I will never be an astronaut. There is nowhere else on earth that could exert such huge physiological strain on my body. By going fully underwater, I have put it under enough pressure that my blood has shifted from my limbs to my central chest cavity. My heart rate has increased. There is up to 700 millilitres more blood in my heart (almost a full blood donation), and so my heart has swelled and must work harder. Because I am breathing under pressure, my veins and arteries have narrowed[1] and my lung capacity is down 10–15 per cent.[2] My body is dealing with a lot. And yet here I float, somewhat ungainly, in this unbreathable ambient medium, and I am at peace.

I am worried about my ears hurting, my calves cramping, and tipping over because I'm bad at buoyancy. Still, I kneel on the sandy seabed while my fellow diving student does exercises and think that here I am, someone terrified of swimming in the ocean, in the ocean and loving it. Even more when I see a fish. What a small ambition, yet it feels immense because all I wanted, and the reason I learned to dive, was this: in the words of the shark expert Eugenie Clark, to be 'looking at a fish and the fish looking at me'.[3] My diving instructor Clément had told me that I would see lots of sea life. He sounded bored by the idea. The other day he saw eight sharks. He loves sharks almost as much as he loves turtles. When he doesn't remember the word for jaw and I tell him to think of *Jaws*, he says that won't work because *Jaws* in French is *The Teeth of the Sea*.

There is little that provides me with fire-bright joy these days, but this does. Not just the fishes, which I record in my diving diary as 'striped fish, black fish, and two sea snails mating', but the sound of the ocean: it crackles and pops. Tiny particles of something or other are floating around, all the time. There are currents. The sea is alive. This is a privilege. On later dives, I see angelfish and starfish, a flute fish, and a puffer fish. I get better at ichthyology with the help of fish books at the dive centre. I see an African striped grunt, a moray eel, a parrot fish, a butterfly fish, bream, an almaco jack, and five stingrays. To me they all seem like a benediction or a present but they are the remnants of what used to live here. It is a like a motorway verge: a pittance and an illusion.

Everywhere there are black spiny sea urchins. I learn the hard way to stay above those. The sea urchins are there because their predators have been fished. At some point, some opportunistic creature will eat all the urchins too if the humans don't eat them first. We are land animals, and we are also the most fearsome and successful marine predators and polluters: underwater I see ribbons and scraps of white, black, and blue plastic like the bag I bought from a street vendor an hour earlier to carry my bananas away in. This plastic has no place floating serenely in the water. I grab as much as I can and am furious that this becomes my focus rather than enjoying the water and the noise and the liberty of it all.

I never panic. I think this is because of my antidepressant, which reduces highs and lows so you float along through life. In fact, living on sertraline is like diving. It produces an emotional state of neutral buoyancy above the sea urchins. If something is wrong, I fix it. Water in my mask: I fix it. Runny nose: I fix it. Clément would approve of this. He had told me that the brain is the organ that uses the most air. The less you stress, the less air you use. 'So I try not to think underwater.' He means that he tries not to think about things that should not be thought about, rather than the things that should, like keeping his students safe. He is a good instructor even with his fatigued zen. He points out sights we should see: a huge ray, or a moray eel slinking along in the sand. He is funny. When he sees the frisky sea slugs he makes the appropriate teenage hand signal involving a hole made with two fingers and another finger. I am entranced by sign language above and below water but at least below water I understand it. At one point he comes to me and starts fiddling with my jacket. I don't know what he's doing, then finally I understand: I am his rubbish bin and he has put a floating plastic bottle in my pocket. We are underwater Wombles.

Because I manage to maintain my buoyancy, I get to go to deeper water: 18 metres of depth at a place called Shark Reef. Don't expect sharks, they say. They see them sometimes although not at Shark Reef. Exciting sightings for the instructors are sharks, rays, moray eels, and turtles. Exciting sightings for me are any damn thing underwater. Clément gets extra animated about a nudibranch, a slug-like mollusc. Once, another instructor was diving and saw a baby electric ray. She knew not to approach it – divers are not supposed to interact with marine life, although some do – but she pointed to it, turned her head to show another diver, and the baby ray touched her and boom!

This water is not Caribbean clear: it is the Bay of Dakar off the coast of Senegal, and the Atlantic Ocean here is murky and mucky. Still, we see a stingray, small fish shoals, a long fish serpenting through the water that my instructor told me later was a flute fish. She swims and points and swims and points.

I touch nothing except accidentally: a bit of rock, an urchin. One of the instructors is a fish whisperer. He strokes sharks' noses and says they come back for more. I can't say who it is because such touching is not permitted. A fellow diver tells me that she gets so angry when divers make puffer fish puff up by poking them. 'It is so horrible,' she says. 'Making them puff up is the equivalent of giving them a heart attack. And these people do it for fun.'

EVERY LAST FISH

This contravenes diving morals. Disturb nothing, leave only bubbles. Unless it is Halloween and you are Clément and dive down to the bay diving site and leave a carved watermelon – pumpkins aren't a thing in Senegal and butternut squash is too hard to carve – a sign saying DANGER DE MORT, and an entire pig costume. There is a night dive every Thursday, and if I were qualified I would do it. I love the trance I fall into when I run in a circle of torchlight on a dark moor; I would do a dark dive to compare. My fears of the ocean, of things that could come out of the dark: off they float. I swim, still clumsily, through this unbreathable world, an uninvited guest in the home of sea creatures, and I wish I could stay.

A nineteenth-century fisherman would not recognize the Bay of Dakar, because for him it would seem empty. I am charmed by everything I see, but it is the dregs. It is what we have left behind after a hundred years of what some people call industrial fishing and what others call mass destruction.

By now we all know about overfishing. No recipe writer or TV chef these days would dare to use the word 'fish' without the word 'sustainable' in front of it. It doesn't matter how carefully you choose your fish, industrial fishing has depleted the oceans so badly that 'eating a tuna roll at a sushi restaurant,' writes the marine ecologist Daniel Pauly, who is sometimes called the grandfather of fisheries science and is by any measure eminent, 'should be considered no more environmentally benign than driving a Hummer or harpooning a manatee.' For every 10 tuna there used to be, there is now one.[4]

Fish for awful statistics about ocean creatures and you will land a giant catch. For every 300 turtles that swam in the Caribbean, there is now one.[5] North Atlantic cod was massacred into near-extinction – in 1992, it was at 1 per cent of historical stock levels – and is only just recovering. The number of glass eels arriving in Europe has diminished by 95 per cent,[6] when the fish used to be so numerous, monasteries demanded rent be paid in them.[7] Sharks are disappearing so fast, there will soon be nothing to be inappropriately scared of. Elasmobranchs (sharks, rays, and chimeras) are the most endangered vertebrates on the planet.

When the National Oceanic and Atmospheric Administration (NOAA) reviewed the state of US fish stocks in 2023, it found that of the populations that fishermen target, '89 percent have a known overfished status'.[8] The Food and Agriculture Organization (FAO) of the United Nations' *The State of World Fisheries and Aquaculture 2024* report concluded that a third of global stocks are overfished and that four-fifths of fish populations are what the FAO calls 'maximally sustainably fished'.[9] Another UN organization, UNCTAD, is starker. 'Nearly 90 per cent of the world's fish stocks are now fully exploited, overexploited or depleted.'[10]

'Overfishing' seems a simple concept. It is actually a technical term that provokes raging debate in scientific journals. The most common calculation uses 'maximum sustainable yield', a complicated way of assessing how many fishes can be safely fished without the population collapsing. I like best the definition of overfishing from the Environmental Defense Fund, as it is 'catching too many fish at once'.[11] Catching too many fishes means that the breeding population is reduced, as well as the number of juveniles, and the population cannot recover.

Catching juvenile fishes is like kicking the legs out from under the future. And still we do it and with such gusto. Fishing efforts multiplied nine times between 1970 and 2008.[12] Four million or so fishing boats now work the seas.[13] They use pelagic longlines, an innocent phrase that means a line of thousands of hooks that can stretch 50 miles, to lure tuna. There are super-trawlers that can retrieve the equivalent weight of 20 busloads of fishes a day, whose nets could trap a plane. Others scrape the seabed with machinery, a process given the prosaic name of bottom trawling. All that effort, and yet we catch fewer fish. Fisheries scientists use up to 40 parameters to assess the health of fish stocks, including 'fish life history (e.g., natural mortality, length and age at maturity, and growth rate), catch (e.g., landings, gear selectivity, and discards), effort (e.g., days fished and number of hooks), and management controls (e.g., fleet allocations and allowable catch)'.[14] Making those assessments sounds like hard work because it is. And we are getting it wrong. A recent paper calculated that a third of what the FAO judges are 'maximally sustainably fished' populations are actually overfished; and the number of collapsed stocks – where the population has dropped to a tenth of its previous maximum – is 85 per cent larger than accepted estimates.[15]

'An early 20th century wooden fishing boat would have caught sixteen times more fish in an hour than its contemporary equivalent,' wrote the academic Chris Armstrong.[16] He calls industrialized fishing 'carnage'. We spend twice as much effort to catch the same number of fishes as we did in the 1950s. You wouldn't know that: catch data has been under-reported for decades.

Many fishing methods don't only capture the fishes they are hunting: what WWF calls 'a staggering amount' of other species is hauled up too. Cetaceans, elasmobranchs, turtles, hundreds of species: it can be anything the fishers cannot sell, do not want, or are not permitted to keep. Sometimes these creatures are landed along with the intended catch. Often they are thrown back into the sea dead or alive, but they usually die anyway. They are called 'by-catch' but a more accurate industry term is 'discards'. Shrimp trawling is often the worst: 90 per cent of what is caught in some fisheries is discarded dead. For every shrimp caught by trawl in the Gulf of Mexico, four times its weight in by-catch is caught too.[17] If all the discards that shrimping entails were included in a retro prawn cocktail, the glass would be the size of a room. The oceanographer Sylvia Earle compared bottom trawling to hunting a hummingbird with a bulldozer.

For Daniel Pauly, the idea that we can still fish sustainably is nonsense because of what is called a shifting baseline. We are content with so little now because we have forgotten we had so much. No one is reading 120-year-old fisheries statistics and understanding the difference between abundance and dregs. 'You're British?' he says in a video call from Canada. 'You have about 5 per cent of what you used to have. At most.' He talks of Scotland's Firth of Forth. 'It was full of huge fish.

And the biomass! The amount of fish you found there was immense. Now there is essentially not one fish. All you find is scampi.'

Seafood is the most valuable edible commodity. It is worth nearly $200 billion in global trade, and this value is rising 10 per cent each year.[18] Between 1 and 3 trillion fishes, probably, are hunted and captured each year,[19] and 100 billion more are raised in aquaculture. Or perhaps it's 10 billion:[20] figures are fishy when the industry prefers to count fishes as masses. (This is why I often use the plural 'fishes', against grammatical convention, because the singular plural noun 'fish' makes it easier to dismiss trillions of creatures as a mass.) Beware when animals are counted in tonnes: the word is used when a number would be more horrifying. The FAO uses numbers for marine mammals, wild-caught crocodiles, and farmed birds and mammals.[21] Only fishes become tonnage.[22] So we know that we slaughter 'only' 80 billion land animals for food.[23] 'Trillion' is so big a number, it is empty. Does this help? It would take you 31,700 years to count to a trillion.[24]

That's just the legal fishing. At least 20 per cent of fishes are caught by illegal, unreported, and unregulated (IUU) fishing. It is illegal because it is done without the right permits or in the wrong places, or with illegal fishing gear (IUU boats reduce the mesh size of their nets so they can catch juvenile fish); it is unreported because more or other fishes are caught than are allowed; and it is unregulated because here are a few things that nation states are not required to reveal if they don't want to: who they give fishing licences to, how many of their vessels fish illegally, and who is the actual owner of a fishing vessel. One in every five fishes imported by Americans is illegally caught,[25] and there is a high chance that the fish Americans buy is not the fish the label says it is: fish is the second most common food to be sold fraudulently, after olive oil. One study of 9,000 commercially sold fishes found that nearly half were mislabelled;[26] another found that the DNA of half of the animals sold as 'snapper' was actually from fishes of 67 other species.[27] This is fish magic, enabled by complex supply chains and murky data: take this Vietnamese catfish, pass it through several fish processors and suppliers, and – abracadabra – now it's a grouper.

The Financial Transparency Coalition found that nearly half of all illegal fishing took place off Africa and that this loses African states nearly $12 billion. Those are numbing numbers. Maybe all we need to know is that one in every five fishes is illegally caught.[28] Every fifth cod, every fifth squid, every fifth everything.

Modern seafood is an illusion of plenty. Most rich countries now import much of their seafood. The UK imports 80 per cent of the seafood it eats;[29] the US 90 per cent. There are no gaps in your fishmonger's display case because of a highly complex global seafood supply system. The gaps are in the ocean.

Worse. 'Peak oil' is a phrase used to indicate the point that global oil production will reach its limit. There is also peak fish, and we are way past it. For Daniel Pauly, we reached peak fish in 1996 exactly. 'Actually in the North Atlantic, where fisheries became industrialized, the peak was achieved in 1975.' Since then, Pauly has calculated, fish catches have declined by one million tons a year. In fact, catch reports – how much fish is being taken – have been severely underestimated for decades. He thinks 'sustainable' is a pointless adjective when it comes to fish populations. 'Everything we do has to be without limit. That is the reason why we have no sustainability. And even if we now had sustainability, it wouldn't work because we need to rebuild the stocks that have been decimated, not be sustainable.' Consider cod. Its population is between 1 and 10 per cent of what it was before. 'So why do we want to maintain that? It should be rebuilt. When we rebuild, the systems heal themselves.'

The tin stinks. It stinks the way it stinks every day when I head downstairs wanting to be asleep but having to give my cat Dora her breakfast. I come to learn that these hours – around 4 a.m. – are the fishing hours. It is when fishermen start work, when fishmongers fetch their stock from fish markets, and when Dora wants to eat fish. The tin holds pink meat, and it is my closest encounter with fish as food although the food is not for me.

Dora's Breakfast

Today's can is skipjack tuna. As usual now I check the label to see if it reveals where the tuna was fished and how, but it could come from the moon, because pet food companies are not obliged to list provenance. There are no cats asking an underpaid server before they order, 'Where is the fish from?' and feeling good for having asked. My cat is a land animal who has never seen the sea yet prefers fish to meat, and I give it to her because I am vegetarian and she can't be.

Tuna must be cheap to become my cat's breakfast. But as we say 'fish' to mean 35,000 species of fishes, 'tuna' means 15 different species. Food websites class them in order of tastiness and how much mercury they are likely to contain. Cheaper is better if you are trying to avoid mercury poisoning, says the US Food and Drug Administration. Heavy metals are more common in fishes higher up the food chain, who swim higher in the ocean and whose flesh is more prized: so skipjack has less mercury than bluefin or yellowfin. Bigeye – given the charming scientific name *Thunnus obesus* and called 'fat tuna' in most languages but not English – is more likely to be contaminated with metal. Skipjack tuna is routinely in the top five of most fished animals, and tuna in general is Europeans' favourite fish. Bluefin tuna is the most lucrative: one once sold for more than $3 million.

And now the ichthyology: if fish were cars, a bluefin tuna would weigh the same as a Volkswagen but have the speed of a Lamborghini. Although it can be 3 metres long and weigh 900 kg, it can swim at 50 miles an hour. If I found the flattest, easiest road race and ran my utmost, I would be eight times slower even though I would have access to far more oxygen than the tuna. A bluefin's movements are impressive laterally and longitudinally: it makes massive vertical dives to find prey and travels across oceans to find suitable waters to spawn in. In fisheries management language, it is a highly migratory fish. To survive temperatures that can range from 3 to 30 degrees Celsius, it keeps its blood warm by swimming and digesting and clever physiology that works like an excellent condensing boiler. The tuna has a *rete mirabile* ('wonderful net') of arteries and veins that can pass heat between each other; the tuna gets warm swimming even in cold water, then can transfer the heat to muscles, viscera, or eyes. The author Oppian, writing 2,000 years ago, called tuna 'the dashing Tunny, most excellent among fishes for spring and speed'.[30] The bluefin tuna is a wondrous animal capable of astonishing physical effort.

So we trap them in pens and shoot them with a shotgun.

Not always. Tuna are caught in many ways. Possibly, they may even be caught in ways that are printed on the canned tuna you buy for lunch: dolphin-friendly, sustainable. They are also caught by large trawlers that use reconnaissance planes to spot the migrating, swarming tuna, then herd them into vast nets that also trap dolphins, young fishes, anything. The most expensive tuna are caught then transferred into 'pens'. They are called farms, pens, or fattening farms, or – considering the highly migratory nature of the fishes they contain – you could also call them prisons. They are essential, according to one fish farming publication, to serve the 'international sushi scene'. One enterprising Spanish tuna pen company provides 'possibly one of the best adventures of your life', by letting paying clients swim among the condemned tuna.[31]

This must be timed right, because once the tuna are fat enough, they must be dispatched and not in a tourist-friendly way. The European Food Safety Authority (EFSA) investigates the welfare of farmed tuna. From its report I learn that tuna are killed underwater by a diver shooting a *lupara*, actually a variation on a Hawaiian harpoon. Or they can be shot from above with a regular shotgun by a shooter standing on a platform. Underwater is better: the cartridge contains less powder so as not to damage the expensive flesh of the fish. 'When the bullet hits the head, it deforms and causes a bigger lesion, but at the same time it ensures an instantaneous death.' Up to 4 per cent of tuna require a second shot. The urgency comes from market pressure: death must occur swiftly 'to prevent lactic acid formation and accumulation in the muscle, because this imparts a metallic taste to tuna flesh when eaten raw as sushi or sashimi'.[32]

The EFSA report is sobering but my imagination is vivid: I see blood in the water, tuna dropping to the bottom of the net, others frantic in their limited space, a diver with a gun pursuing a distressed tuna to get a second kill shot before the tuna is taken from the water and its head is removed with a chainsaw. Anyway, I don't need my imagination when there is YouTube. Underwater, a diver readies the *lupara*. Massive fishes swim past him: they are Atlantic bluefin tuna – *Thunnus thynnus* – and they are majestic. The scene looks serene, as the underwater generally does in film, because you can't hear the creaks and the sonar and the crackles and the fishes singing, and because we land animals have no idea how noisy the ocean is now from natural and human-derived racket. (I find a paper that has explored whether farmed tuna get stressed from wind farm noise, but that seems less crucial than the stress of getting a shot in the head.) I can't hear the sound of the *lupara* firing, as the diver takes aim at a tuna at random and shoots it in the head.

I look at my cat's breakfast differently now.

'You don't know what you're missing.' I have been told this over cod, pollock, salmon, prawns, scampi, lobster, oysters, and hake, and over fish breaded, fried, stewed, and raw. I have heard it said over my mother's famous fish pie or the salmon starter at endless dinners or the steaming heat of a freshly battered fish from a fish and chip shop. I have seen people hold up their forks and wave them at me and say yet again that I should try it, just once. Everyone who tells me this thinks they are the first person to say it, as if the thousandth repetition will be different and suddenly I will like the taste of fish.

I know what I'm missing because once it was in my mouth and I spat it out. Fish. The staple food of billions, and I don't like it. The fight began in my primary school dining room. Round tables with low child-height stools, attached by steel trusses to the table so that you thought that when the children left at the end of the day, perhaps the table woke itself up and began to spin like a merry-go-round. Lunch was fish fingers: neon orange sticks with white middles of some fish or other. Three-quarters of Britons get their first taste of fish from a fish finger.[33] I hated them. But this was 1975 and my dislike was irrelevant to the dinner ladies who policed our food intake. Eat it. Eat it, or you can't leave the table. Eat it, or you will have no pudding. Eat it.

I ate it. At the age of five, I learned about taste buds and I moved the fish finger unchewed to the safe side of my mouth, away from my tongue. I left it there until I could leave the room and spit it down the toilet. And that was the last time I consumed fish.

That is a lie. I don't eat fishes for food but I would have to make an effort to escape consuming them. I feed fish to my cat. I use a non-fish omega-3 supplement (produced from algae, which is where fishes get their omega-3), but I drink fish when I have beer or wine because it is used in the processing; I apply fish to my mouth because it is used in lipstick; and I scatter fish on my garden because it is in fertilizer (that's an easy one: the sack is branded 'fish blood and bone'). I may have been medically treated with fish: chitosan – a sugar from shellfish outer skeletons – is used to treat obesity, kidney failure, and high blood pressure. Fish is in biodiesel, other pharmaceuticals, natural pigments, and cosmetics. I can buy marshmallows made with ground fish bones; or shoes, wallets, and handbags made from stingray leather, described by a leathercraft site as 'the most unforgiving of the exotics',[34] because it is hard to cut and hard to stitch. There is no escaping fishes.

I know my distaste for fish makes me an outlier. The world wants more and more of it. In the 1960s, the average person ate 9.9 kg of fish a year, and now it is 20.5 kg.[35] The biggest fish-eaters are the people of the Maldives: they consume a massive 80 kg a year.[36] The contrarians are Japan. Twenty years ago the average Japanese person ate 40 kg of fish a year, now it's 21 kg. In 2011, the Japanese began to eat more meat than fish.[37] Japan is going its own way as usual and getting fatter because of it.

Most rich countries have a boring fish palate. The most commonly fished fish worldwide is anchovy, although much of that catch becomes fishmeal. The five fish most frequently eaten in the UK are cod, haddock, salmon, tuna, and prawns,[38] and the first two are usually battered and served with chips. Americans go nuts for shrimp.[39] Russians like oily fish: mackerel most, then herring. The Dutch like herring so much that a plate of *Hollandse nieuwe*, new season herring, was offered as a bribe for people to turn up for their Covid vaccination.[40] Salmon is the favourite of sushi nations. In 2021, the Taiwanese took their sushi love to an extreme. A newspaper headline read, 'TAIWAN OFFICIAL URGES PEOPLE TO STOP CHANGING THEIR NAME TO "SALMON"'. The lure was a bottomless sushi buffet for half a dozen people if your ID card included the character for 'salmon'. People changed their names to 'Salmon Fried Rice' and 'Explosive Good-Looking Salmon'.[41]

A third of the planet gets almost a third of its protein from fishes. A fish can provide a human with not just omega-3, but fatty acids, carbohydrates, protein, lipids, eicosapentaenoic acid (essential for growth and brain health), and linoleic acid. There are micronutrients such as vitamins and minerals that aid oxygen transport, hormone and metabolism regulation, immunity, and growth. It's so good for you, maybe I should eat it.

A campaign by the ocean conservation group Oceana was titled 'Save the Oceans, Feed the World'. What animal protein, it wrote, 'requires no fresh water, produces little carbon dioxide, doesn't require arable land and provides healthy protein at a cost per pound lower than beef, chicken, lamb or pork, making it accessible to the world's poor? The answer: wild fish.'[42] Bigging up seafood is standard in modern food security thinking. By 2050, our fish consumption is predicted to double.[43]

Where will it come from?

In 2010, the marine biologist Victoria Braithwaite published a groundbreaking book. She called it *Do Fish Feel Pain?*,[44] and I wonder at the doubt when it should have been a statement. The UK's Animal Welfare (Sentience) Act, which established in law that fishes are sentient, became law only in 2022.[45] The US has no federal laws establishing sentience, and in essence all animals are property.

Fishes are animals that look weird and they are wet, but they are alive and they have lives that we take from them because we are hungry or because they are tasty. And how do we take them? We stamp on their heads; we suffocate them so their eyes pop out; we stick hooks in their mouths and pretend their mouths are numb. We axe them in the head, although the axe is called a 'gaffe' and is a brutal way to get a fish on board a boat. Stab and heave. As Jonathan Safran Foer wrote in *Eating Animals*, 'no reader of this book would tolerate someone swinging a pickaxe at a dog's face'.[46]

Fishes are animals that are rarely included in animal welfare laws, whether the animals are pets, farmed, or wild. Is this because humans assume an animal has fur and legs? I come to call fishes 'portable food units' (PFUs) and wonder why no fisher I meet, not even the most sensitive and thoughtful, sees them as anything else. The answer is that they can't afford to. You cannot think fishes are sentient individuals who laugh and sing and have friends when you leave them gasping on the deck for hours or slice them open while they are still living. You must forget that cuttlefish are as clever as Labradors when you drown them in the trawl or let them be cannibalized in traps. I meet fishermen who tell me how canny tuna are, and how artful the bass, then rip a hook from a mackerel's mouth and leave it bleeding, and respond, when I ask how long a mackerel takes to die, with, 'I don't know. I've never thought about it. Not long?'

They can do this because for a creature we have eaten for thousands of years, the fish is still a mystery. There are dozens of academic journals about every aspect of fishes and fishing from the *Journal of Fish Biology* to *Aquacultural Engineering* to *Marine Drugs*; marine biology departments are stuffed with students intending to spend their career studying the ocean's inhabitants; David Attenborough and *Deadliest Catch* are prime time. And fishes are still misunderstood among all that apparent scrutiny. I'm concluding this from a few things: how little they still appear in animal welfare law; how inhumanely they are killed and captured, and how little that is questioned; and what we call them.

I don't mean their common names. Those are the bee's knees. The Dolly Varden, the common bully. Alewife, bonytail, and the convict cichlid; gravel diver, monkeyface prickleback, and the pineapplefish. The wobbegong, slimehead, and the requiem shark. The midshipman fish, the warty angler, the upside-down catfish, and (my favourite) the mouth almighty. These are called common names, and they can vary between fishing ports, never mind countries. The language of fisheries management is more indicative of the place of fishes in our consciousness. Not 'fishes', 'populations', or 'death rate', but 'fisheries', 'stocks', and 'yield'. All these words assess fishes but only how they serve humans.

That is why it has taken so long for us to know things about fishes that we should have known before. They fart, they communicate, they feel (they definitely feel pain), they love, they nurture. Sharks have friends.[47] Minnows shout at one another.[48] In Bali, where all industry stops on Nyepi, an annual day of silence, researchers recently recorded four fish choruses in one night.[49] Fish sing!

Most people encounter fishes as food or pets, or flopping onto a dock. They are thought pea-brained, if they are thought sentient at all, and not just moving food units that will be a fish finger or a Friday fish supper. The World Trade Organization classifies fish as an industrial product, not an agricultural one. Fishes, brake pads, same difference.[50] Fishes are still for most humans nothing more than Samuel Coleridge's 'slimy things' in the water.[51]

When WWF wanted to campaign about the mass slaughter of blue-fin tuna, it put a panda mask on a fish with the slogan, 'Would you care more if I was a panda?'[52] In a letter to the journal *Animal Sentience*, two academics wrote that 'public support for conserving and protecting specific species concerns the "cute and cuddly" or "beautiful but deadly" animals (e.g., pandas, penguins, tigers, polar bears, gorillas, or whales)'.[53] They pointed to the honeybee as an example of making previously unloved species beloved. The honeybee is not in danger of extinction (although other bees are), yet now everyone wants to protect pollinators. It is a successful conservation story. This isn't the first example of bees and fishes coming together. The court of appeal in Sacramento was recently asked to decide 'whether the bumble bee, a terrestrial invertebrate, falls within the definition of fish'.[54] To enjoy a sensation of falling through space, I recommend reading this judgment, as it concludes that 'although the term fish is colloquially and commonly understood to refer to aquatic species', the word 'fish' is also 'a term of art' and can therefore include bees.

To find a true friend of fish, we must go back 2,000 years to an ancient Turk named Oppian, who wrote *Halieutica*, a 3,000-line poem about fish, to please the Emperor Severus.[55] Oppian decided to write of 'all that inhabit the watery flood and where each dwells, their mating in the waters and their birth, the life of fishes, their hates, their loves, their wiles, and the crafty devices of the cunning fisher's art – even all that men have devised against the baffling fishes.'[56]

And what fishes these are. The ravenous and shameless dogfish, the impetuous black race – for their dark hearts – of the tuna, the home-loving lobster, who 'holds in his heart a love exceeding and unspeakable for his own lair and he never leaves it willingly'.[57] We humans have no grounds to feel unique in our territorial obsessions. Fishes have homes too.

In Oppian's oceans, octopuses love olive trees, and when they smell one, they will leave the water and climb it because they have 'caught a passion for its grey-green foliage'.[58] The 'sargues' are more fun, because they go mad for goats. 'Goats they yearn for and they rejoice exceedingly in the mountain-dwelling beasts, even though they belong themselves to the sea.' Picture a herd brought to the sea for a bathe by their goatherds, besieged by herds of sea bream, who 'leap joyfully on the terraces by the sea and fawn upon the horned company and lick them and crowd about them with many a gambol'.[59]

Why do I love *Halieutica* when it is so daft? Why do I not exult in my superior scientific knowledge that octopuses do not climb olive trees and sea bream prefer shrimp to goats? Because for Oppian, the fishes have as much personality as the fishermen. He elevates them to equal standing with humans, and that is rare and fine. They are fishes, not tonnage. Their characteristics and foibles and wiles are described or devised with such care and dexterity that I want it all to be true, including the part where fishermen catch fish using onions. Oppian knows that fishing is a brutal job, but he also writes that fishermen can be devious and cruel. Cheer then for the fishes who 'not only against one another employ cunning wit and deceitful craft but often also they deceive even the wise fishermen themselves and escape from the might of hooks and from the belly of the trawl when already caught in them, and outrun the wits of men, outdoing them in craft, and become a grief to fishermen'.[60]

The world's 38 million fishermen still have plenty of grief.[61] One NGO estimates that 100,000 die on the job annually, and although this figure is triple previous estimates, it is probably inadequate. Fishers are called the last hunters and, like hunters, they have never known what they will catch. The warming ocean means fish populations are moving away from expected fishing grounds. There are too many boats, not enough fishes, and not enough fishers. Rich countries face the problem of what is called a 'greying fleet'. The average age of fishers is 50 and their children don't want to fish. Poorer countries have plenty of fishers and nothing left to fish.

Fish farms will save us. Already more than half the fish we consume is farmed, and aquaculture production has been growing faster than wild capture since the 1980s.[62] For the second time in history, humans have turned from hunting to farming, only this time with crops that swim. At least that is the rhetoric and the hope. Some researchers talk of 'aquaculture overoptimism' and point out that unless we figure out how to keep wild fish in the sea, aquaculture production would have to be three times the average currently predicted level to meet demand.[63] Farmed fish are still fed wild fish; for now, one cannot exist without the other, and the other is disappearing.

As land is polluted and fought over, as water rights on land trigger more conflict, fishes will become more central to our food security. Can they survive us? In 2024, the Java stingaree, a ray, became the first marine fish to be declared extinct because of human activity. This was, said one conservation biologist, 'a tipping point for marine diversity'.[64]

When two waves meet at sea, the result is called interference. Various things can happen, and the most arresting and rare is a cross sea, where the waves form neat boxes like an ocean chessboard. A cross sea is highly perilous because of the riptides it generates. But it is still a bewitching rendition of collision, of forces that meet with unknown outcome.

A collision is coming between humans and fishes. I don't mean that the poor fishes will rise up, that the fish will grab the gaffe and axe the fisherman in the face for a change. But that all assumptions about our relationship with the creatures of the water are in the air. We assume they will always be there in the ocean. We assume that what is on our plates will be 'natural', as if strychnine is not natural, as if farmed fishes don't get antibiotics and chemical lice treatments.

I wrote a book on shipping and thought that was complicated. I hadn't yet encountered fishing. Shipping uses flags of convenience that make it easy to slide away from regulation and prosecution; so does fishing. All ships must carry a permanent ID number; fishing boats aren't required to.

Fishing is meant to be controlled by regional fisheries management organizations (RFMOs), which are supposed to set fishing efforts at a sustainable level. Maybe they try. But they are voluntary, and states that disagree with the catch levels can leave and keep fishing as and where they please. RFMOs usually focus on a single species, which makes no sense in an ocean ecosystem. Or they set catches at levels that suit the greediest bully in the organization. The workings of RFMOs are not easily accessible: when I asked to observe a one-hour Zoom meeting, I was told I could, for $500. A marine biologist has called the work of RFMOs 'managed extinction'.[65] There are sustainable fisheries – Alaskan pollock, orange roughy – but more are unsustainable. The fish scales are tipping the wrong way.

I am a nauseous vegetarian. Until recently I would hold my nose while walking past fishmongers' stalls. Vomiting while I watch fish be hooked, trapped, bludgeoned, suffocated, and sliced is not my idea of fun. But I want to bring alive that dead fish on your plate, and also the person who fetched it from the sea or the salmon farm. So I have wandered above and below water (salt not fresh); I have met fishermen and fisherwomen and liked them, and watched fishes die. I have swallowed seasickness pills and vomited nonetheless because my body does not like the ocean's currents, and still I have gone back to sea. I have done it because I want to know where we are with fishes, where fishes are with us, and where all of us are headed.

© Rose George

Watching the nets

2

A Man in His Boat

Must be something wrong with anyone who wants to go fishing.

DAVE, FISHING HERITAGE CENTER, NEW BEDFORD

The sea looks nothing like trouble but my stomach is heaving. My hosts are unaffected. 'It happens to anyone,' they say. 'More food for lobsters.' They offer me water; they are gentle with me, then they get back to collecting crustaceans.

The *Nimrod* is small as crabbers go. The crew talk wistfully of bigger boats that can collect thousands of pots. Today we will gather only 190, all laid out in strings of creels – 'creels' and 'pots' are used interchangeably – on the seabed. We are only on the second string and I'm cold and queasy and wishing it to end. This is not because of the company: my hosts are Rex Harrison, his son Alexander, and Rex's son-in-law Dave. Rex was named after a florin, not a movie actor (when he was born, coins bore the word 'Rex' for 'king'). With his weathered skin and white blond hair curling past his collar, he seems a clear descendant of the Norsemen who came over the sea to northern towns like Filey and left their mark on local faces and in dialects: 'garth' for a garden, or 'watter' for water. I don't tell Rex any of this fancy, because although we have only just met he would tell me he has the tan of a fisherman, not a Norseman. Nothing other than an office perk because his office is the water and the wind of the bay.

I first encountered Filey and fishing when I read an account in a local newspaper of 'the Filey Few'.[1] Half a dozen fishers still went out, the article said, when years ago there were dozens. By 'going out', they did not mean to the high seas. Most Filey fishermen now either went out potting, gathering pots of lobsters and crabs from the seabed, or they fished for sea trout in the few months a year that they were allowed to, laying their nets to catch the fish as they travelled along the coast. The Filey fishers now barely leave their bay, and in decent weather they can always see what is happening back on the beach. I tell Rex that I haven't been to Filey since I was a child and he says, 'It hasn't changed much. The only thing that has changed is that the fishing community has gone. When you came there, this hill [he means the Coble Landing, a cobbled slope that leads to the beach] would have been full of boats. Cod and haddock in the winter; lobsters, crabs, and sea trout in the summer. Now some of the fishermen have moved to Scarborough. They've got bigger boats, and they're fishing for lobster all year, but it can't sustain it much longer. They're actually working the grounds that traditionally Filey boats worked, so we've got no say in it now.'

Rex fishes sea trout as well as shellfish, and the sea trout season is limited to between April and July. Rex is permitted to fish crabs and lobsters all year although he chooses not to. Sea trout are related to salmon, and this makes them lucrative because they taste so good, they are so few, and the season is so short. This is not why Rex and his fellow fishermen still refer to their 'salmon boat' and 'salmoning'. Until recently they were allowed to catch wild salmon, but now they are not. Also their sea trout season has been reduced by a month, supposedly to reduce the risk of by-catching salmon, which often hang out with sea trout. Rex was quoted in many newspaper articles on the injustice and stupidity of this decision: he and his fellow Filey Few had gathered 25,000 signatures from locals in favour of the fishermen, but the petition was as useful as most. Rex mutters about 'all them salmon' being caught by rich river anglers and rightly he can't see why the anglers aren't prevented from killing salmon who have come to spawn, when they take so many and the Filey Few took so few.

My march to our meeting point on the Coble Landing takes me past the Royal National Lifeboat Institution (RNLI) shop and lifeboat station, past the shuttered shops that will soon display plastic spades and buckets and sunglasses and fishing nets. I pass a kiosk called Sugar Snax and 'the best fish and chip shop in the world'. ('No,' said Rex later, 'they cook chips from frozen, they're rubbish.') I had been instructed to arrive at 5 a.m. sharp and to look for the *Nimrod*. That was easy when there were hardly any boats to choose from, yet there was no one in sight. They were waiting on the other side of the landing, hidden behind an SUV. Three men, all old: one in a neon orange beanie hat, one with a smile, and one Rex. 'You're late,' says orange beanie. 'No, I'm not,' I say sharply, my humour hard to find at that hour. He wants to have fun at my expense even so and probably because of it. He looks at the sky and says, 'It's going to rain.'

'No, it's not, I've checked the forecast.'

'I didn't say where.'

That was actually funny. I call him 'orange beanie' because he is not introduced to me. Two more men arrive and aren't introduced either. There are more important things to be getting on with. Everyone gets busy, while I loiter politely, unneeded among industry, a fish out of water.

Finally Rex tells me to get in the boat, and I climb up onto the tyre that serves as a step and lug myself over. It is not elegant but better than Rex's belly flop over the side, which makes the crew laugh as apparently it always does because he always does it. Rex had asked me to come early to accompany him crab and lobster potting, 'so you can see all of what we do'.

Now I'm in the *Nimrod* but not at sea. We are still parked on the Coble Landing because Filey has no harbour. Boats do not bob here, waiting to be lifted by water. Instead, the *Nimrod* is attached to a tractor, and we trundle down the slope to the beach then into the sea. This is why the coble became the most common boat along England's east coast: its flat-bottomed form made it easier to launch from a beach. Its pedigree, like that of the men who crew these boats, is Viking. Cobles are extremely quick and deceptively hard to handle.

Rex doesn't call his salmon boat a coble but a salmon boat. It is not wooden-built, nor is it row-powered. Its shape is of a coble even if it has an engine. Rex's coble can be called by many other names. A small-scale fishing vessel; an Under Ten; an open boat. It is the Filey version of the African pirogue, the Asian sampan, the Cornish punt, the dory, the buss, the Madagascan lakana, or the Indian clinker. *Nimrod* is his bigger boat, an Under Ten, as under-ten-metre boats are known. Under Tens make up more than two-thirds of the British fishing fleets but only hold 2 per cent of the quota, the government-issued permit to catch certain fish.[2] Under Tens are the boats seen on graciously lit Sunday evening programmes in pretty harbours. Most fishes are caught out on the high and deep seas by boats that are not bucolic but brutal in their size and their capacity, but half of the world's 4 million fishing vessels have no engines. These figures are best guesses, along with the number of fishes our greed has left in the ocean. Some authoritative numbers retreat like surf when you poke at how they were calculated.

It is bizarre to be on a boat travelling over sand. It feels strangely regal. What it is not is streamlined. The tractor takes us to the water's edge, then three of the four men aboard jump into the water and push the boat off. Now I see why they looked suspiciously at my wellington boots: theirs are not the traditional laced sea-boots of before, which were often kept unlaced so they could be kicked off when a man fell overboard, but they are thigh-high waders. Theirs are not overwhelmed by lapping waves as mine are. All my kit is from the land: my waterproofs from my hobby of hill running; my boots from my allotment. I am making do.

Suddenly we are afloat, then moving, then at sea.

The other men on the *Nimrod* are Alexander, Dave, and Chris. We will drop off Chris, an employee, at the salmon boat anchored out in the bay, where he will sit for hours watching the nets. Dave is married to Rex's daughter Nicola, who works for the Coastguard. She goes sea swimming at 5 a.m. before work and I like the sound of her. Sea people. The local church flies a brass fish above its tower as a weathervane, and plenty of its gravestones bear an engraving of a boat because plenty of its men have died at sea. Filey had 190 fishing cobles in the late nineteenth century, 17 in 1984, and 7 in 2001.[3] Rex is a remnant: Alexander is now fishing only because he lost his job on cargo ships during the pandemic. Dave works on the oil rigs and is a fisherman in his three weeks ashore. Orange beanie and his mate are retired and get up this early for fun.

Rex's father was a fisherman and his grandfather too. Rex first went fishing 'in a nappy'. His school was up on the cliff where the town is, and he remembers the boom of the maroons – rockets – that were fired from the Coble Landing to signal that the lifeboat was needed. He was allowed out of school when the maroons went off and sprinted down the Ravine, a long steep hill, with the other fisher boys. 'The teachers in them days, because it was a tighter community with lots of fishermen, they knew you had to go.' Rex's wife, Adrienne, does not fish. This makes her unusual in Filey's history, so intermingled and insular that a common old joke about Filey folk was that 'if you kick one, they all limp'. A common Filey tale tells of an old woman who decided to travel by train for the first time in her life. 'Where are you going?' asked the booking clerk. 'Mind your business,' she retorted. 'Just like all your family. I'll walk.'[4]

The marriage registers of St Oswald show that of the marriages recorded at the church between 1813 and 1908, 'all 34 grooms were fishermen, and 27 of the brides were daughters of fishermen'.[5] There was so much mingling that families gave each other surnames for names, and men were known better by their nicknames. In a folder in a local museum, I find pages of these, printed out and with handwritten annotations. Snottynose Dalton, Backstrap Charlie, Jocksie. Diddy, Tom Pop, Bighead. Dickie Bunny, Snow-white, and Slosher. Should you have asked for the best fisherman in Filey by his name – Richard Cammish Jenkinson – you would have received blank looks. He was instead Dicky Hoy and the man with the top coble.

Filey's tourism logo is now 'Filey makes you smiley'. It is a popular place to visit, its bay still glorious and its clay cliffs stunning like the rest of this majestic Jurassic coast. The *Sunday Times* recently decided Filey was 'one of the six seaside towns everyone will be talking about' because of 'its laid-back atmosphere and lengthy sweep of golden sands'.[6] It did not mention Filey's fishermen even though the presence of fishing boats in picturesque coastal villages is thought to add serious money to tourism revenue. A search for 'fish' on Filey's official tourism site returns several images of the 'fishtive' Christmas tree – made from lobster creels, topped with a fish, and erected every Christmas on the Coble Landing – and an article wondering if the withdrawal of salmon licences meant the end of fishing in Filey.

Not yet, not quite.

It is cold, and I am glad of my warm clothes and that I have extra in my bag. All the men are in what fishermen call 'bibs'. They're not really bibs. They're also called oilskins and they aren't those either. They are neoprene dungarees that have retained an old name from habit. Under those, the men wear fleeces, not woollen jumpers. Traditional fishing sweaters were knitted so closely as to be a second skin and were worn next to the body for better waterproofing. To prevent the chafing caused by wind and seawater, men wore silk at the wrists and mufflers at the neck. The current version of this is layering and thermals. We all wear hats because the breeze becomes a wind the further we go, even if that is not far. This inshore fishing is almost at-shore fishing. The pots are all in or just beyond the mouth of Filey's wide and grand bay. Each of the Filey Few has his own sites for his strings of pots, marked by buoys. Alexander and Dave get ready at the side of the boat, side by side because now they will work as a team, while Rex is at the small wheelhouse, watching the water and the monitor and keeping the boat where it should be. He watches the water and I watch him.

The winch brings up the buoy, then rope, then the first creel, and then it becomes like television for me, entrancing to my morning state: Alex hauls up the pot and opens it for the catch that is crawling inside. Dave puts his hand into the bait funnel, which holds the rotten fish heads that lure in the prey, and throws the used bait into the sea while grabbing another handful from a plastic tub nearby. Quickly, they assess the size of each crab or lobster and either turn to put them

A Man in His Boat

in a bucket or hurl them back into the sea with such force I wonder whether the lobsters spin all the way to the bottom, dazed. Half of the catch is thrown back. I ask the first of my dumb questions. Doesn't it harm the crabs to be flung in like that? 'Nay, they can swim.'

I thought lobsters were pink. I thought sea trout were brown. I've got everything wrong. The lobsters are black with a radiant shimmering blue on their tails. Sea trout are pink. Female lobsters are called hens and even the internet cannot tell me why.

The pots are all hauled and stacked carefully on deck by Dave. I wonder how I am going to stand another two hours of this. Yet here we go again. I pass the time by timing how long it takes them: one hauls the pot, opens the bait cage, flings the old bait into the sea, while the other takes out the live catch, reaches to get new bait, fills the bait cage, and turns to stack the creel with the others.

Seven seconds.

At some point, Dave and Alexander turn to assess and sort the catch, and I feel that I can ask questions. What are you doing? Why? There are seven signs to look for on a lobster. They can't be landed if they are pregnant, and you know that by turning the lobster over and looking for eggs, which are called berries. Hens make berries. Briefly, my brain considers the peculiar ways the human mind works and why it would give a crustacean the same name as a chicken and also not call an egg an egg. Pregnant hens get thrown back. The lobsters also have to be a certain size because, as every fisherman knows, you don't go for juniors because then you won't have seniors to catch. A lobster with a V notched in the tail is a breeding hen and is thrown back. 'Who notches them?' 'Other fishermen.' The shell has to be hard: a soft shell means a young lobster or a female which has just moulted – shed their shell – and is growing a new one. Soft shells are lobbed overboard. All this heedless handling is hard to take and I knew it would be. Soon I stop looking at the crabs in the tub, at how they wave their legs in shock when another crab is lobbed on top of them, then they quieten down, resigned. They have a long haul ahead: they will either be put in a creel and kept in the sea for a day or two, or they will be placed under carpet that is always kept damp, put in a truck, and driven along a road for eight miles or so to Scarborough. Even then they will not die.

I pity the crabs. I pity the lobsters. I know this is no place for pity, and I knew I would feel this way and that there would be no way of showing it. It would be impolite. I also have pity for me because I feel my stomach start to lurch. The waves are not big but the swell is enough. The end to this nausea is more than a hundred creels away. Now we steam to the next line of pots, the conversation dies, unable to compete with the engine, then the boat stops, and Rex turns his head to watch Dave and Alexander and I try to ignore my stomach. Rex has just had a sandwich. It is lunchtime for someone who got up at night. 'What's in your sandwich, Rex?' 'Lamb,' he says, his mouth full. 'Last night's barbecue.'

A bird craps on me, then again. I ask the men how often they get shat on and then apologize. I have read that swearing is not allowed on most boats, along with other things such as women. I apologize for what my sex may be doing to the fishing, and Rex shrugs. He has no patience with the notion that women on boats are bad luck: if they catch no fish today then it will be because of the weather. If birds crap on me, it's because we are a couple of miles from Bempton Cliffs, a shrieking mass of seabirds that is a protected bird colony, and wonderful, and deafening.

I'm thinking of superstitions and birds and anything I can to distract myself. To ease everything, I chat to the men when I can, between their hauling. Crabs will eat anything: they are the pigs of the sea. The strangest thing they have caught is a £5 note, 40 miles out, and a tennis ball that was given to Rex's dog. They mention the sunken Roman pier, now reduced to boulders on the seabed, which crustaceans like. There are rumours that Filey's seabed also holds old fridges and gas stoves, thrown there in the nineteenth and twentieth centuries by Filey coble men who hated the new trawlers that were coming from Scarborough with their huge nets that were dragged along the seabed, stealing their fish. Anything to trip up the trawl.

A Man in His Boat

In Aberdeen, hostile fishermen stoned the first trawler that attempted to come into harbour.[7] In Filey and Scarborough, feelings ran high enough that a sea commission was convened, with investigators interviewing all sorts of men involved in fishing or with fishermen. The Reverend Arthur Petit, 'incumbent of the Iron Church', was asked what fishermen's complaints were. 'I find their constant complaint is about trawling, because in their opinion it injures the spawn of the fish.'[8] In 1863, when Reverend Petit gave his evidence, he estimated that there were '400 or 500' men engaged in fishing. Fishermen thought the trawlers spoiled the fish by dragging them along the seabed in their trawl nets. They were sure that the trawlers cut through their own lines. They knew that 'before they came there used to be plenty of fish and now there is very little of any kind'.

There are no trawlers now in Filey Bay. The finfish industry (finfish have fins and gills; crustaceans don't) has disappeared from the east coast. These inshore fishermen, as they are categorized, rarely go out of the bay, hardly ever beyond the 12-mile limit that is considered British territorial waters. I can look around, gingerly, and see the extent of Filey's current fishing industry: Chris, sitting watch on the salmon boat, and orange beanie, sitting in another salmon boat with his mate. That's it.

Witnesses to the 1863 enquiry fished for 'cod, ling, haddock, skate, halibut, turbot, soles, plaice, thornback and herring'. The only finfish on the *Nimrod* are those caught accidentally and kept for Rex's supper, and the rotted cod heads used as bait. Rex eats fish at least twice a week; Alex 'could take it or leave it', and Dave doesn't like it at all. I ask them what different fish taste like and they look blank. 'They all have different tastes. I can't turn around and say they taste like fish, can I?' Rex doesn't like cod or bass but anything else is fine. 'I could eat a whole sea trout. With a salad. Then I'd fall asleep for a week.'

The time has passed, the contents of my stomach have stayed put, and we are nearly done with the pots. Dawn walkers are now out on the Brigg, the long finger of rocks and boulders that stretches out to the sea and that has wrecked plenty a ship.

Rex and Alexander talk wistfully of buying a catamaran. It would be more comfortable and they could double the number of pots. This boat, says Alexander, 'is tiny'. Alexander's father made him go to nautical college, and Rex's father did the same to him. Rex was 15 when he went to 'sea school' in Gravesend for three months. 'Then I went to sea in the Merchant Navy until I was 20-odd. He would not let me come home to go fishing but I still enjoyed being away. Then I came home and went fishing.' When Rex took Alexander up to Scarborough for his interview for nautical college, Alexander was already fishing on a boat out of Scarborough. Father and son rarely fish on the same boat and not just in the Harrison family. 'Just the way it's done,' says Rex lightly, although it is actually grim insurance. You may lose one but if they are not on the same boat it is unlikely you'll lose both. In 1925, the *Research*, a steam trawler, was heading into Bridlington in a blizzard. Three huge waves later the trawler was sunk. At once Fanny Jenkinson lost her husband, two sons, and two sons-in-law. Another woman, mother of two men lost, never locked her door again in case they came home. On Fanny Jenkinson's gravestone in St Oswald's churchyard, the epitaph reads, 'She suffered much but murmured not.'[9] Sea people.

In that Scarborough hotel the college interviewer asked young Alexander what he had been doing the day before. Rex tells the story. 'He says, "I was at sea." The fella said, was it bad weather, and Alexander said, "Yes, they sent the lifeboat for us." The man was astonished. "You were at sea in that weather? Were you frightened?" "No," said Alexander. "It's all I know."' And the man immediately offered him a place. I ask Rex if he has ever been scared at sea. 'No. There's times when it gets close and it all goes wrong and if a wave comes, especially in the old open coble boats, they might be half full of water. But you just get on with it, get the pump on, bail it out, and carry on. I don't think about it. It's just part of the job. If you worry about it, you wouldn't go to sea so much, then you wouldn't earn nowt. It's not bravery. It's your life and you don't know no more.'

Rex bought his first boat from a fisherman who had a heart attack, and he was skipper of his own boat by the age of 24. The family bought the *Nimrod* for £6,000 ($7,400), then the fishing licence cost £19,000 ($24,000). A catamaran, he says, costs 'not that much', only £50,000 to £60,000 ($62,000–$75,000).

'These are the best pots,' says Rex. We have turned back to face Filey. 'We know what's in them.' I am slow to catch on. I need hot coffee and some carbohydrates so of course my questions are stupid. 'Why are they the best?' 'Because they've got Wednesday's catch in.' On a hot day, it is not worth the risk of killing the catch by driving 8 miles to Scarborough in a hot van, so the crabs and lobsters are caged and stored in creels underwater.

I'm numb now to levity. The shore is right there, and we are almost done. Get it over with. I can be home for lunch.

'We're not done,' says Rex. We're going back out in the salmon boat. My heart sinks because the reporter has to keep going while the woman wants to make the churning stop. Then I realize that we are going ashore to have breakfast first. Alexander and Dave are taking the catch to Scarborough; we beach, jump into the water, leave the boat for the tractor to deal with, and I walk up the landing to sprint, as much as wellies allow, to the toilets, only just open because the day has barely begun although mine feels halfway through. Everything feels better with coffee and food. Rex has a cheese toastie, though he hasn't had one for years, and asks for a tea because, he tells the server, 'I've fallen out with your coffee, it's useless.' The Coble Landing is busy now. I tell Rex that I really like Filey and we look out to the bay. 'You can't beat it,' he says. 'It's so peaceful. I come down here every day even in the winter for a cup of tea or coffee. They put a couple of heaters under there [he points to the awning] and I have a tea. It's best when it's quiet like this.'

We are joined by a young lad. He looks like a student, and his name is also Alex. Alex is Rex's granddaughter's boyfriend, and this is his summer job. I look at the sparkling sea and the long looming finger of rock that is Filey Brigg reaching out to sea and think, it beats an Amazon warehouse. The Brigg is where Filey's young women used to go to collect bait for their menfolk. Filey calls mussels 'flithers', and so they were the flither girls. They were hardy, fabulous women who wore pantaloons for ease of descending the cliffs, and stout boots. The Victorian poet and 'man of letters' Arthur Munby became fascinated by the flither girls of this coast, leaving a photographic record of them and writing about them in his diaries. Of Sally Mainprize of Flamborough, he wrote 'spitting on her hands (vulgar creature!), and rubbing them together, she firmly grasped the rope and stepped over the edge'. The girls could descend with one hand on the rope and one on their basket; the men needed two hands. Sally would stop and sit on a ledge 'at the corner of the jutting craig', and 'whistle to the waves'. I gaze over to the Brigg in the hope of seeing a vulgar creature descending one-handed on a rope, whistling at the water. There is no one there.

It is time for salmoning. We will relieve Chris, who is already out in the boat, and we will stay out until 3 p.m., fish depending. I feel like a tourist attraction because tourists are looking at us and, though I have not yet bought oilskins to look the part, I am with obvious fishermen. Tourists approach Rex and Alex all the time. 'My favourite is when they ask you if sea trout is cod or haddock.' One of the defences for the fishing industry's continued subsidies and disproportionate place in the nation's consciousness – it actually contributes a tiny 0.03 percent of GDP, about the same as the lawnmower industry – is that tourists are drawn to coastal villages with bobbing boats in the harbour.[10]

The salmon boat is pink and named *Australynn*. Rex doesn't know who *Australynn* was: that was how the boat was named when he bought her and so that is how she is still named. It is bad luck to change a boat's name. The cheese and bread and coffee have given me enough wherewithal that I have asked what to expect. We will be back ashore in daylight. They used to fish overnight, but that was disallowed several years ago. They also used to leave their nets and go off somewhere else, but that is also disallowed. They sit and they watch the nets. This kind of fishing is called static gillnetting: a net hangs in the water, anchored on the surface by buoys. Other gill nets can drift. Some can be miles long. This one is modest but the fishing is still taxing.

'You don't read?'

'No. I'm not one for reading, and you have to watch the nets.'

'What about audiobooks?'

'No.'

'What do you do all day?'

'Watch the nets.'

Alex says, 'Come on a bad day. It was snowing in April.' People pay money for this kind of enforced contemplation: it's a meditation retreat at sea. Rex would probably laugh at that. It's just the job.

If a fish is caught, it moves the net, so Rex and Alex must watch for the agitation. On a good day, they can catch 30 or more sea trout. On a bad day, none. Rex knows the tides and he knows the waters, but he can't fathom the fishes. After a while, when we have caught nothing, he mutters about the west wind having buggered things up.

'Why does the westerly matter?'

'They don't like it in the bay.'

'Why not?'

'If I knew that, I'd be a millionaire.'

It is a peculiar fact of fishing that it is by nature an extractive and murderous industry, yet small boat fishermen like Rex are also conservationists in a way. They know their environment and understand it. Like the fishermen who stoned that Aberdeen trawler, they are conservative in the other way. Change is difficult when you have used the same gear and techniques as your father and grandfather and they still work. I am asking Rex something, and he starts to answer then leaps up, saying to Alex, 'Aw shit, get your gloves on.' The radio continues to play cheery pop music but the atmosphere is now tense. I can't get up in case I disturb the boat so I watch Rex's back. He gets Alex to motor over to a part of the net, and they huddle over the side fiddling with something. Fiddle, fiddle, curse, fiddle, then a bird is released from the net and Rex launches it into the water. 'A razorbill,' he says. 'See his beak? There you go, it'll do that [skitter along the water] until it buggers off. The faster we are the better, so they don't get caught in the net so much. Normally they're out quicker than that. You need to witness that because people say we are liars.'

I'm slightly stunned by all this: the sudden change from nothing to hurry, a bird in the net, the urgency, and the vehemence. Liars? Who says that? 'Twitchers.'

Twitchers are birdwatchers, and what just happened is in fact the result of many years of cooperation among twitchers and conservationists and Filey's fishermen. What just happened, all five minutes of panic and huddle, then that razorbill scudding across the water like a skimming stone, is a conservation success story.

The Royal Society for the Protection of Birds (RSPB) has a video on YouTube.[11] It features three men in a boat: Rex Harrison, a bearded Scot named Rory Crawford, and another man from the Environment Agency. Before I met Rex, and before I saw this video, I read various official reports about the Filey conservation success, which were as dry as most reports are. Rex in the video cuts through all that in perfect Rex style. 'When this first started, I'd have chucked all this lot off the back of the boat.'

'This' was seabird by-catch. Fishers are always catching things that are not their intended target. And their nets, and the fishes in them, also attract the ocean's other residents, those who gather in the hope of scraps, who soar and screech and cry and caw. It comes under the catch-all category of by-catch but technically a drowned seabird – they usually die – is a discard. Seabird deaths by fishing can happen in many ways: they can get tangled in lines or nets. They can get speared by hooks when they come in to get the food that is put there as bait. Their lungs soon fill with water; they drown. BirdLife International thinks seabirds are one of the most threatened groups of birds in the world and notes that their numbers have declined by 70 per cent in 50 years. Other numbers are also grim: 200,000 birds are killed by fishing every year. 'That's 23 every hour. More than one every 3 minutes. That means that every time you brush your teeth, by the time you're done, by-catch has killed yet another seabird.' Fishing, the report continues, is the single greatest threat that birds face at sea.[12] Longline fisheries – long lines of hooks just below the surface – are particularly bad; so are surface gill nets and anything that uses a floating device to keep the hooks near the water's surface.

Discarded drowned seabirds were part of fishing life and for many years Rex just accepted it, despite living in sight of Bempton Cliffs, an RSPB reserve. 'Because we'd been doing it for years, it seemed normal.' Then the bird people came.

Rory Crawford works for BirdLife International. He started with albatrosses, a heartbreakingly common victim of fishing, and saw countries such as South Africa and Namibia reduce albatross deaths by more than 90 per cent, simply by changing their gear or using tools such as bird-scaring lines (lines that carry streamers) or lines weighted so that the bait is deeper and out of reach; or lines that are set at night when few seabirds forage.[13] 'It's thousands of birds saved per year,' says Crawford. 'It's dead effective, this stuff.' Much of conservation, he knew, was 'toiling for years and you just keep seeing depressing stuff happening'. He knew also that campaigning to save albatrosses was 'glory-hunting', like supporting Manchester United. Albatrosses are romantic, and they are easy to fundraise about. It's harder to raise money to save the wee Filey kittiwake.

Birds were being caught in the Filey nets because the nets were doing what they were meant to do: they were invisible. On the video, Rex explains: 'What happens is when we put the salmon nets in the sea, the feed, what they're feeding on, gets trapped next to the net and goes into a feed ball. And the problem was that the birds were going through the feed ball and getting stuck in a monofilament net at the back.' Rex's first encounter with a conservationist official was not a success. 'There was an RSPB conservationist team,' says Rory. 'Generally their work is focused on grouse moors and birds of prey persecution.' They used the same techniques of monitoring and observing from Filey's headland and the Brigg, and the fishermen noticed, and they were furious. It was 'very antagonistic,' says Rory, and he talks of 'deep valleys' that can open up between communities. Fishing can be insular and, in the views of fishermen, subject to enough surveillance already.

When Rory first came to Coble Landing, 'they were all stood there with their arms folded'. He was apprehensive because he knew fishing and he knew fishermen. 'You don't always get a contact like Rex. Sometimes the person who is the most willing to reach out is not the most influential person. But it was great having Rex because he is that person. He almost is Filey, like the human manifestation of it.' He follows this by modifying it because he knows Rex. 'Without sounding too over the top and Rex would definitely balk at that.' Rex remembers the icebreaking clearly. 'They were wearing these puffed-up lifejackets, those foam ones. I just looked at them and started to laugh and that was it.'

It sounds modest, but Rex and his fellow fishermen did something extraordinary. They decided to see if they could improve their gear. If the nets were invisible, then why not make them visible so that birds could see them? 'It was someone who said, why don't you try the pigeon netting that's on the roofs? So my net is all that.' Rex immediately switched from a monofilament to a heavier, darker net. In the first year, all seven fishermen of Filey, as there were then, reduced seabird deaths by 90 per cent. Such a figure lures people with clipboards as surely as the sand eels draw the salmon.

It took another couple of years to find the perfect material, and along the way they experimented with floating corks on top of the net. 'You know when you see guillemots and razorbills in the sea, they're all white, so we thought white corks would attract them. Then we tried grey corks.' But it wasn't until they devised a net that wouldn't catch fish that things improved for the birds. 'It sounds daft, doesn't it, a fishing net that doesn't catch anything.' But as well as changing the material of the net, they also changed its shape. Now, it's a J-shape, and the fishes are funnelled into the loop of the J, while most of the net – the long wall in gillnetting terms, the stem of the J – makes it easier to release any stuck birds.

The bird I watch being freed is the first bird trapped this year. 'It's still early. Towards the end of next month, when gannets come, you can sit here and watch 5,000 feeding. That's summat to see.' He waves at one point to the sea and the cliffs. 'My office. Weather's different every day, you're outside every day. You have to get up at stupid o'clock in the morning, but you get to see stuff that people never see. Bottlenose dolphins prancing round in the middle of the bay. We've started seeing them more off here. And it gives you freedom, we can pick and choose where we want to go. And we have the benefit that we don't have to go, but for some unknown reason we feel like we have to go.'

There is pleasure in Rex's voice when he talks of seabirds and dolphins, and of the seaweed that he picks from the sea and eats there and then, that tastes just like lettuce. I like to hear him talk about things he likes, from his wife coming out in the salmon boat then swimming back to shore, or the 'right fun' he has with Dave and Alexander sometimes. Some things though are hard to articulate, like why he left a job that many young men would consider perfect – travelling the world on a cargo ship – to come back to this bay, these Jurassic cliffs, this small parochial town that faces out to sea and immensity. Rex can manage no more than the entirely sufficient, 'I just wanted to go fishing.'

There is no pleasure in his voice when he talks of seals. On another trip out in the salmon boat, he is again mid-sentence when he shouts, 'BASTARD SEAL', startling me with the loudness of it. For several years now the job of watching the nets for fishes has included watching for seals. Rex hates them. 'We used to have just one, all summer long. Now there's thousands, around the back of the Brigg there's a colony and another colony of 400 to 500 halfway to Whitby. So they're eating all the fish. They upset the eco-balance. There's too many.' He can't and won't kill them. 'I don't want them wiped out, but I do think they need culling. When I was a kid we'd catch sea urchins and take them home. They have thorns on them like rosebush thorns, and the seals are even eating those. There's a fish we call the lumpsucker, they have little nails, some are bright orange. They're beautiful-looking fish, but there's none left. The net used to be full of them. You'd get them out, alive. They've gone.'

Jerry Percy is a former fisherman who used to run NUTFA, the New Under Ten's Fishermen's Association. When I asked him what fishermen were calling him about, he was quick to reply. 'No fish.' He has just got off the phone with the latest caller, an inshore fisherman like Rex who had been out. 'No fish.' NUTFA has now closed; Jerry is getting old and funding was never easy. In an interview about the closure, he pointed out that four out of five British fishing boats are small-scale. They are Rex, not giant factory ships with onboard processing. In his eyes, it is the giants that get the subsidies and the quota, and government promises about safeguarding coastal communities are empty when it won't give low-impact fishermen – the Rexes – any quota or peace.

'We use 17 times more effort to catch the same amount of fish as we were catching in 1901. The whole panoply of modern fishing gear, if you compare it to when I was a boy fisherman, it was nothing compared with what we have now in terms of materials, electronics, sonar, radar, fish-finders, ground discrimination sounders.' Most of fishing's best weapons were adopted from military ones. For Paul Greenberg, fish expert and author of *Four Fish*, 'World War II was a tremendous incentive to arm ourselves in a war against fish.' A modern trawler will catch fewer fish than one in 1923 because of what Daniel Pauly calls the 'depletion-expansion' model. 'Thanks to their onboard technology, trawlers and other industrial vessels could fish anywhere in the world, in deep or shallow waters and far from coastlines, and in conditions from tropical to polar . . . Previous obstacles to fishing – depth, distance, ice cover, and inclement weather – could now be overcome. As a consequence, essentially all fish resources in the world are being fished.'[14]

Rex, with his modified nets that are now world-renowned, has no sonar or radar. Rex has many names in policy documents. He is an artisanal or small-scale or low-impact or Under Ten fisherman. For Rex, he is just a fisherman. The only electronic device in the salmon boat is a radio playing Heart FM. Rex is his own fish-finder. And he is finding no fish. He sits on the edge of the boat, his legs dangling over the water. He looks at home. He looks resigned.

'It's going to be one of those hauls.'

A Man in His Boat

'A bad one?'

'Aye. Wind and tide. We got 20 last week and our lad only got one.'

Fisheries management organizations, scientists, governments: everyone tries to figure out how many fishes to fish and why fish stocks go up and down. Rex has an answer to that.

'It's the way it goes.'

He knows that times are harder now. He remembers when there were 100 fish a day in his nets, and that wasn't long ago. It is easy to fall into nostalgia in fishing. In shipping, reminiscing about better days is called swinging the lantern. Rex does not indulge in this, with the same commitment with which he refuses to discuss politics although politics governs his working life. He uses the lantern to look back at what was and also to go forward. Now his problems are the seals, but also the ocean, the big European trawlers that are allowed to fish in what fishermen think are British waters, disturbing the nature of things, in the same way that trawlers in the 1970s turned the peaks and troughs of Filey Bay's seabed into a flat sandscape. The trawlers would come from Scarborough and Bridlington when it was dark, says Rex. 'We could go out at 4 o'clock in the morning from here, we used to steam down the bay with no lights on because it's easier to see where you're going in the dark in winter. As soon as we got to the fishing grounds, we turned our lights on and the trawlers turned theirs on and it was like Blackpool Illuminations some mornings.'

The last time I visit Rex, it is October. He and Alexander are standing at the coffee kiosk, closed for now, looking out to sea. I look too because they have an intensity that seems to contain urgency. I imagine a swimmer in trouble or a mass of seals. No one in Filey does winter fishing any more so it can't be a fishing issue. The weather makes putting out crab pots in winter too risky: Rex could lose many thousands of pounds of pots in a bad enough storm. Opposite us is the new catamaran, newly bought second-hand, unable to fish until April. Alexander and Dave have come down only to look at it. It is Alexander's future: £50,000 worth of optimism that fishing will continue; that the small-scale fisherman can still fish among the massive trawlers and the shadier operators of the industry. Alexander looks startled when I ask what they are looking at with such ferocity. 'Nothing! The sea.' It would not do to be gazing at a boat as if it were jewels.

Later, Rex and I sit on two chairs under the cafe awning ('I'll book a table,' he had said. By this he means, 'I'll put summat on it and they'll leave me be.') He has allotted me an hour, although I had asked to come to hear his life story, because he has promised to take his wife shopping. We sit here where he comes every day, winter or summer, because he is drawn to it, and look through the drizzle at Filey Bay, its grand sweep between the Brigg and Bempton Cliffs. The view is impressive even on a dull autumn day. 'When you think about it,' says Rex, his eyes turned either to the water or to the expensive new boat that will keep the Harrisons going even longer. 'When we were all younger and the whole town was involved in fishing . . .'

He stops.

'It's sad really. I never thought I'd be one of the last.'

Real fish and chips

3

Fast Fish Food

It is quite likely that fish-and-chips, art-silk stockings, tinned salmon, cut-price chocolate . . . the movies, the radio, strong tea, and the Football Pools have between them averted revolution.

GEORGE ORWELL, THE ROAD TO WIGAN PIER[1]

On a quiet street in an agreeable Leeds suburb, half a mile from pleasant woods and trails, there is a low grey building with a small and humble sign. This reads: NATIONAL FEDERATION OF FISH FRIERS. Other than the sign, the building gives nothing away. Recently, my running club did a training session here – the streets are usefully hilly – and the National Federation of Fish Friers (NFFF) was a topic of conversation because it intrigued everyone. Why such a small building for such an important association? Everyone loves fish and chips. It is so beloved in the national culture that Winston Churchill refused to ration it in the Second World War.[2] And why is it here? We're as far from the sea as you can be in England. It's not even in the city centre. Why can't you smell fish? What do they do in there?

People have theories. It's a KGB front. There can't possibly be anything inside it. One clubmate says he has read that there is a training centre for fish friers and he's sure that it is behind that plain door. I disbelieve him but he's right, and so here I am again just after eight on a Monday morning, learning how to make fish and chips. I arrive during a discussion of a potato's dry matter, starch, and sugar content, take a seat on the back table among piles of demonstration takeaway boxes and coffee cups, and start to pay attention to the instructor, a veteran fish frier from York named David Miller. A wet potato loses money because it uses too much oil. Frying should never crackle: that's water in the fat. Slicing a potato's eyes off with a knife not a peeler loses money because you lose chunks not peelings.

Times have been hard for fish and chips. Andrew Crook, the ebullient president of the NFFF, has a list of problems. Fish and chip shops were allowed to stay open during Covid lockdowns, but Crook closed his. 'I said, right, so would you send your wife and your daughter to work in a busy fish and chip shop facing a thousand customers on a Friday night with nothing to protect them?'

Now the vexation is cost. Everything costs more, from the energy needed to heat the friers (they are called 'ranges'), to the potatoes, which are now £30 ($37) a bag when they were £5 ($6) two years ago – the highest they have cost since 1976 – to the fish, which the industry used to mostly get from Russia but now mostly can't. The NFFF has lost a tenth of its members over two years as far as Crook knows, though he guesses it might be more by how many copies of the NFFF magazine are returned to sender. Yet people still want to buy into the industry. David Miller often trains people who are buying a fish and chip shop and who know nothing about oil, fat, or potatoes, or fish. When I tell him I don't like to eat fish, he is not ruffled. 'I've met people buying shops who are allergic to fish.'

Today's course is not for apprentices but established friers: two are Tabby and John from Blackpool, which David describes as 'possibly the busiest fish and chip town in the country'. When David asks them how many chippies there are in Blackpool, they say 'a lot', then calculate that there are four – no, five, one just opened – on one corner on one street. They used to get their fish 'frozen at sea' from Russian factory trawlers. Haddock mostly but also cod. After the invasion of Ukraine, prices began to fluctuate and then kept increasing. The UK government also imposed tariffs on Russian fish, although these could be swerved easily enough by Russian ships landing their catch in Norway or elsewhere. Now the friers get haddock – fresh, not frozen at sea – from somewhere north of Scotland. 'Rock something?' says John. 'Rockall?' That's it. They think it's much better.

David is puzzled by their account. He exhorts us all to say 'fresh frozen' rather than 'frozen at sea' because that word 'fresh' is money, even though 'fresh' fish is many days dead given the length of trawling trips (there is no legal requirement for what 'fresh' actually means when it comes to fish). It's the haddock he can't understand. He has a geographically precise understanding of the UK's fish and chips preferences: 'Leeds into the start of the Pennines. Drop into maybe Doncaster or Sheffield. Then it spreads out to the east coast. So Scarborough, Bridlington, all that way to Hull and up in Scotland.' That is haddock country. Beneath that, people eat cod. So Blackpool's choice of haddock is baffling. Tabby and John don't have much response to his bemusement other than it's not affecting sales; hardly anyone asks for cod and they are run off their feet frying haddocks. When they later tell Dave that their fish and chips only costs £8.99 ($11), and children under 10 eat free, he is amazed, because another of his phrases is 'fish and chips is not a cheap meal'. In his shop in a middle-class village outside York, a fish supper costs between £10.30 and £11.70 ($13–$14), depending on the size of the fish and the weight of the chips.

When the Blackpool chippies tried to sell pollock instead, customers didn't like it. 'It's more opaque,' says Tabby. 'They were sending it back saying it wasn't cooked, but it was.' In his book on fish and chips and the British working class,[3] John K. Walton records a much more colourful map of regional fish preferences from the early twentieth century. In fish and chips, as in everything else, the nation's palate has dulled. Bury used to like sprag, Radcliffe preferred hake. 'Hake was also coming to the fore in Manchester and Ashton-under-Lyne, but in Birmingham and Walsall it was merely tolerated as a cheaper substitute when plaice was expensive. Haddock was first choice over much of the industrial West Riding but in Nottingham nobody would touch it, however cheap the price.' Londoners were broad in their tastes, liking haddock, plaice, and cod but also skate and rock salmon (actually a smooth hound shark also known as a dogfish). Hake was most popular in Birmingham and Cardiff, where a local speciality called an 'elongater' was also much in demand. It was said to be 'something between a ling and a conger' but was so obscure that, a civil servant gleefully remarked, 'the Fisheries Division could not tell us how to spell it'.[4]

Preferences are as firm as a scallop shell, yet efforts have still been made throughout fish and chip history to forcibly change them. This year, Andrew Crook visited the West Coast of the United States to source possible new fish stocks. The two main candidates are Pacific haddock and rockfish. Pacific haddock isn't perfect for their needs. 'It has a very fragile skin, so they've got to do it as soon as they catch it and they don't really process on vessels over there.' He's leaning more to rockfish because there's loads of it (for now) and it's not Russian. 'It's never going to replace haddock, but it might take off a bit of the pressure.' In 1935, writes Walton, an NFFF official admitted to the Seafish Commission that customers were unknowingly eating dogfish. During the First World War, friers took 'what they could get: first ling, then catfish, sold under the euphemism "Scotch hake"'. Coalfish became standard in war years, though both its name and its nickname 'Black Jack' were appropriate: its dark flesh was disliked 'as well as its name' (the actual flavour and texture were widely agreed to be excellent).[5]

There are 10,500 fish and chip shops in the UK,[6] compared to 1,200 McDonald's and 840 Kentucky Fried Chicken establishments. The British and visiting population eat 382 million meals from fish and chip shops every year. A quarter of people visit a fish and chip shop once a week – I like this language on the NFFF website, as if people are calling in to pay their respects to the shop, not just leaving with a fish supper and a jumbo sausage with mushy peas – and three-quarters visit at least once a year.[7] The fish and chips industry has always had a high opinion of its importance and influence despite public snobbery towards it: in 1927, when the *Daily Herald* estimated that 30 million portions of fish and chips were served every week in Great Britain, a reporter wrote that 'should fashion veer suddenly, and fish restaurants pass into the limbo of things forgotten, one in every five of our trawlers would probably lie idle'.[8]

Fish and chips have travelled beyond the United Kingdom, although not always successfully. A chippy called Malins, set up in Tokyo in 2014, claims to be Japan's first.[9] It is named after one of the supposed pioneers of fish and chips, the East Ender Joseph Malin, who served fish and chips from a shop in Bow in 1860.[10] (Other parts of the UK contest this and offer their own fish and chip firsts. They're still arguing about it.) A journalist writing in the *Aberdeen Press and Journal* in 1927 about this 'minor trade that is thriving' was both precise and vague about his choice of origin story, writing that 'some forty years ago the pioneer of the fried fish industry in this country made his appearance in the form of a stout man with a barrow drawn by a donkey. He commenced operations in an English manufacturing town, and led his donkey throughout the streets, calling out his wares in a stentorian voice the while.'[11] Who was the stout man? How do you keep fried fish warm in a barrow? The journalist does not spoil his story by revealing his source.

Malins' menu is predictable, offering fish and chips, fish cake and chips, and also salmon and chips for less than £8 ($10). In Paris, the fish and chip shop Mersea gets applause for the best and cleverest pun, and none for its mushy peas, because they are lentils in cider vinegar. Non, merci.

The United States would be the promised land for fish and chip purveyors – all those people with all those big appetites – except the United States has resisted fish and chips as firmly as its people have mostly resisted eating fish. In 1909, Americans were eating about 70 kilograms (kg) of red meat and chicken and only 5 kg of seafood a year.[12] By 2021, they were managing to consume 9 kg of fish.[13] In his book *American Seafood*, the chef and fisherman Barton Seaver presents plenty of history to explain this food preference. The Puritans thought fish inferior to meat, even when the tides washed in tons of the stuff for free and for no effort. William Bradford, governor of Plymouth Colony in 1623, wrote, 'If the land afford you bread, and the sea yield you fish, rest you a while contented, God will one day grant you better fare.'[14]

This distaste was exacerbated by the great American agrarian dream: the true American had a farm and land and land animals. He did not have a fish pond or a fishing boat. Some immigrants came from fish-eating cultures, but they were working class and therefore so was fish. As for its association with Catholics, that didn't mesh with the Puritan first families either. All in all, fishes were despised and feared: they were messy, smelly, and bony. They were common. Fishes have always had a place in religion: Catholics were urged to avoid eating warm-blooded animals on a Friday to honour the Crucifixion. They could have eaten snakes or newts but fishes were easier. Then in 1966, following a decree by Pope Paul VI, American Catholics were released from the obligations of fish Fridays.[15] Seafood consumption dropped even lower.

Two things changed that: a man called Clarence and the coronavirus. In 1912, Clarence Birdseye visited the Canadian province of Labrador while working for the US Department of Agriculture. There, he noticed that the Inuit custom of deep freezing freshly caught fish in ice produced something after thawing that was still tasty and still noticeably fish. Birdseye didn't understand why, but the key was the temperature. A lower temperature means proteins in the fish's flesh are not attacked by microbes. The cells keep their shape. The fish tastes like fish.[16]

The Inuit were way ahead of existing technology in the US, where freezing techniques used higher temperatures and produced mush. Birdseye experimented: American housewives were offered 'fish bricks' that they were supposed to slice when they needed.[17] But American housewives did not yet have refrigerators and American shops did not have freezers. He tried frozen herring blocks, which he called 'herring savouries'.[18] They did not catch on. Then he switched to cod, turned his bricks into batons, invented the fish stick, and the lunches of billions of school children would never be the same again.

There are compelling videos on the internet showing how fish sticks are made. One is in German because the German port town of Bremerhaven has two fish stick factories that produce 350,000 fish sticks a week. The video is called 'Fish with Corners'. In the Iglo factory (Iglo is the European brand name of Birdseye), each block must be the right size within three millimetres, then it is cut four times to make 378 sticks. These are 'enrobed' in batter, an august verb for a mundane factory machine process, then enveloped in breadcrumb, flash-fried, and there is your orange fish stick.[19]

For me and anyone else who grew up in the UK, it is a fish finger. When Birdseye launched the baton of fish with the orange breadcrumb in the UK, the batons were almost called Cod Pieces.[20] The women on the production line, the story goes, chose the name 'fish finger'. About 50 years later, a survey of British primary school children found that a third thought fish fingers were made from chickens or pigs;[21] and a fifth of youngsters questioned in another poll thought fish fingers were fingers.[22]

Fish finger facts I enjoy: different nations prefer different colours of crumb. The Italians like theirs to be yellower; the Britons want garish orange. The colour, according to fish finger people, comes from paprika and turmeric.[23] The fish used to be cod but now is often Alaskan pollock. It is hard to diversify from such a beloved fast food. Birdseye launched a salmon fish finger in 2009 that was withdrawn only six years later.[24]

Why was the fish finger so successful? Because it was fish that was not fiddly. A 1955 advertisement in Britain promised, 'No bones, no waste, no smell, no fuss.'[25] The same year, an American commercial was certain that 'even folks who never cared for fish love this new taste thrill'.[26]

In the UK, the fish finger got a mascot only in 1967. He was Captain Birdseye, a rugged sea captain with a bushy white beard who spoke with the West Country accent that is somehow associated with both seafaring and pirates. Captain Birdseye was avuncular and charming and spent a lot of time hanging out with children on various boats and ships, offering them fish fingers 'at the captain's table'. This was genius pester power: children began to demand fish fingers, and parents began to buy them. Captain Birdseye was international, although he became Captain Iglo across most of Europe. The beard definitely helped his fame, as Graeme Rigby writes in *The Herripedia*. 'Angus Watson had brought the bearded fisherman in his sou'wester to the Skippers brand; Raskin's had a bearded Jewish grocer for their Schmaltz Herring; Captain Birdseye arrived in 1967.'[27] In the US, Gorton's followed the template, with its bearded Gorton's Fisherman with yellow sou'wester and oilskins, although TV versions were clean-shaven and, being American, had perfect teeth. Gorton's hasn't yet changed its mascot even though the Fisherman and Gorton's are now as American as sashimi, as Gorton's was sold to the Japanese seafood multinational Nippon Suisan Kaisha in 2001. (This fact is not included in the Our Story section on Gorton's website.[28]) American or Japanese, Gorton's Fisherman is doing the job. During the pandemic, fish stick sales rose 30 per cent in the first six months of 2020.[29]

When Birdseye retired Captain Birdseye in 1971 – advertising this with an obituary in *The Times* – and tried to replace the actor John Hewer with a land-based family, the company had to resurrect the captain three years later because of uproar. They eventually replaced Hewer with another bearded man, then a stubbled one. I notice that news of Captain Birdseye's removal was filed under 'Weird News' in one newspaper. Then I find that Gorton's has launched campaigns featuring 'Mer-bros' alongside their stoic Fisherman. Buff men with fishtails exhort millennials to eat more frozen fish products. I think the ads look like their creator was on crack, but sales increased by 6 per cent.[30] Millennials with less time are buying more frozen fish, even when it is bright orange, crunchy, and advertised by mermen doing bicep curls with shells.

In the prep kitchen, we are ready for fish but David is mortified. The fish he fetches from the fridge is yellow and awful looking. 'It's not good. I would reject that.' He seems truly upset. NFFF gets donations from companies and can't refuse what it is given. All this testing uses a lot of food. The NFFF approached the city council to ask if they could give the food from training courses to a homeless kitchen, but they were refused on the grounds of health and safety. Someone might get food poisoning. 'At least we tried.'

This yellow haddock does not look promising but it will serve the purpose, which is to show how to cut fish to make the most money. Filleting is a skill and David has it: he is deft with the knife and makes it look simple, slicing a bit here, removing a bone there, making three bits of fish – a regular at 150 grams, a mini fish that could be sold to children, and a bit to be used for fishcakes – out of one. I watch him and remember a young man I saw at a freezing-cold fish warehouse in Grimsby at 5 a.m. The young man wore headphones and danced while he sliced. He was the best filleter there and a worthy descendant of the Grimsby filleters, who were so good they were dispatched to Iceland on lucrative short-term contracts.

David slices and slices; haddock first, then some slightly less yellow cod. The cod needs to be deboned. In the past Tabby and John employed a staff member to stand for eight hours a day cutting, slicing, deboning. 'Obviously they got a bit bored. And their hands started to freeze after two hours.' Now the fish comes already boned, in a better state for frying, and they have saved on the fish-boner's wages and hands. Today's fish is easier on the mouth, not like in 1928, when *Fish Frier Review* columnist Sand Dab wrote that 'our customers in the old days . . . must have approached their fish and chip supper in the same manner that one would expect a fox terrier to approach a hedgehog'.[31]

Today is about self-improvement tips and what sports psychologists call marginal gains. It is a luxury for friers to be here with the space and time and material to test. And the tips are based on either good practice or science. The filleted fish is turned over once filleted because otherwise the fat evaporates and it loses protein. David 'tucks to bed' the prepared fish on a tray and drapes the tray with a plastic film. 'People laugh, but I'm creating a microclimate and that makes a difference.'

*

EVERY LAST FISH

When I was a child, I made a fish and chip shop in my bedroom. I cut the chips out of paper, oblong after oblong, and also the fish. I think I charged for it. I have no memory of ever eating real fish and chips and now will never do so. Now here I am in the NFFF frying kitchen and I am delighted, because it's a fake fish and chip shop. There is a board with prices on it, and everything costs nothing. There are takeaway boxes and bottles of vinegar neatly stacked on a fridge. The other requirement in a chippy is an abundance of carbonated sugared soft drinks, or 'pop'. (Here, water that comes out of the tap, supplied by the local authority or corporation, is 'corporation pop'.) We reminisce about childhood pop. Cream soda, shandy, dandelion and burdock. I ask David why you can always find dandelion and burdock in a fish shop, and he has no idea. 'You just do.'

Possibly the carbonated sugar cuts through the grease in the fish or the chips. Possibly there aren't enough calories already in the fish supper. It is hot with the range switched on and this talk of pop makes us thirsty. David sends out for cold drinks and meanwhile we swelter and learn. I've been wanting chips for two hours now and finally they are coming. We will test batches. It is a competition between potatoes: the rumbled and the pre-prepped (a tautology that means the potatoes come prepared and cut and only need to be fried). Each batch is cooked for a different length of time in the two pans of oil, one heated to 140 degrees Celsius, the other to 160. First the chips are blanched. David points out that blanching actually means to cook through without colouring. To blanch chips at the NFFF is to cook them – 'it's beef dripping, sorry, Rose' – for six minutes at 140 degrees, then for two more minutes at 160.

He drops the first batch in and the oil crackles. 'That's what you don't want.' He serves some of the first batch, which are edible. I've had similar and paid money for them. I've had worse and paid money for them. He puts some of the batch back in for another minute. 'Look at that,' he says, when they come out. A totally different chip. He's right: they look better. I do not question the fact that I'm standing here earnestly judging the outward appearance of a deep-fried potato. It matters when chips are the most profitable item for any fish and chip shop.

Now we turn to batter. It must be made with ice-cold water and should never be left overnight because the raising agents will start to work and you don't want that. You want a batter that clings to the fish like a desperate lover, so when you cut through the cooked batter, there like a treasure is the appealing flaky white flesh of the cod or haddock or Black Jack or plaice. David favours a pre-made batter mix and has tips about water temperature, whisking techniques (from back to front to get the air in, and with pizzazz), and most importantly the flow rate. Batter is what mostly goes wrong, says David, because nobody measures the thickness by passing it through a flow cup. This looks a lot like a menstrual cup but I keep that to myself.

'First we're going to cook an awful fish,' says David. 'Really bad.' An awful fish will demonstrate how good a good one is. I don't join in the tasting now. Anyway, the difference is visible in the cling: the thicker batter is claggy and leaves a gap around the fish. The best is the batter that flows through the cup in 25 seconds. By now the testing table is laden with fish but even so David asks the class, 'Do you want me to do you a fish? I can do you a fish.' He gets polite refusals then honesty and hungry stomachs prevail and everyone gets fish and chips for lunch, or just chips, and we take them into the classroom to eat, so soon it smells more like a fish and chip shop than the fake fish and chip shop.

For fish friers, 'the Friday tea' is an institution and an ordeal. David jokes during the training course that it is how you know a trainee frier has got it or not: if they survive the Friday tea, they will be OK. Friday is the busiest day for any fish and chip shop. It is a cultural habit that sits on top of old religious belief. It is also to do with paydays and tradition. The French called fish days *les jours maigres* (the thin days), and meat days were fat ones.

David invites me to come to his shop in Haxby for the Friday shift: 4 p.m. to 9 p.m., a closing time late enough for all comers, not late enough for pub drunks. Haxby is a genteel commuter village north of the genteel city of York. At first I drive past Miller's, David's shop, because its front is hidden by a huge blue parasol and outside tables. You don't get that in most chippies. I don't think David likes the word 'chippy'. It is too reductive for a business that serves halloumi fries and gluten-free fish and chips, and tonight has chicken loaded fries on special as well as a baked haddock in garlic butter.

Inside, he begins to introduce me to the staff. The first surprise is how young they are and then how middle-class they seem. Fish and chips, no matter how refined the shop, is still a trade, and educated young people today are rarely directed towards the trades. I used to marvel in Italy that being a barista was seen as a noble career and wished we didn't scorn skilled trades in the UK. Here there is still class and snobbery and both have now reared up like a salmon, challenged and found wanting by these smart young people content to work in an industry still perceived as common.

Tonight Lewis is frying, Ellie is front of house, Scott is in the small in-house restaurant, and May is 'on the door', which means she is nowhere near the door but taking orders from 'walk-ins'. After the pandemic changed habits, walk-ins are the minority. Most of Miller's business is online and mostly delivery, so there is, according to the website, 'a carbon-neutral delivery network so that . . . your hunger will find salvation'. The carbon-neutral delivery network is actually eleven lads on bicycles.

The friers – two on fish and one on chips – stand at the back. John, bearded, older, and tattooed, is on screens today. This means he prepares the fish for the friers by checking the orders that come through on two screens. The bigger is for online orders, the smaller is more urgent because 'the people are standing right there'. Ruth is on chips tonight. She has come to Miller's from a fish and chip shop in Staithes on the Yorkshire coast. Before that, she worked in fine dining for 10 years. This is a surprising career move: fish shops are no longer regulated under the 'offensive trades' section of the Public Health Act[32] but no chef would see them as a step up from a fine dining kitchen. 'Is that usual, Ruth?' She laughs. 'No.' During the pandemic she had to leave her chef job, then started working in fish and chips in her home town. 'I enjoy the routine. I like knowing what I'm doing all the time.' Friday tea or not, it is less stressful. Ruth's confidence is as charming as the fact that she's wearing false eyelashes to fry chips.

Three people have already come into the shop although it's not yet 4 p.m. For the next four hours, the energy ebbs and flows, with long periods of busyness and urgency – not panic – punctuated by moments where the energy drops and, although it is never quiet, it is quieter. More staff arrive. Some are former riders who have 'expressed an interest in frying', as David describes it. I understand this. I have an interest in frying, or in bagging chips, but although I have been loaned a Miller's T-shirt and one of the hip denim aprons, there is no room for me: at the shop's full complement, there are 12 staff in the front of house, doing sides (mushy peas, gravy, curry sauce), bagging, scooping the chips from the chip well, putting salt and vinegar into sachets. Three more are pot-washers and kitchen helpers. So I move around the place, trying not to get in the way, choosing different loitering points, trying not to slip as the floor gets greasier. The gluten-free frier is a good leaning spot, as today it is off duty and only operational on Thursdays. David will not budge from using beef dripping but otherwise he can be flexible. Jack, a farmer's son who is firm that you can't run a farm with only one tractor, is a pot-washer, the person in charge of cooking mushy peas, and the operator of the vegan frier. Miller's offers vegan chips fried in vegetable oil, as well as vegan chicken strips and a jackfruit 'fish'.

I note down the communications between front and back of house, between friers and packers, between Sam and John.

Have you done Tundra yet?

Angela's gone now.

Get rid of Jackie.

I really hope that these comments don't relate to the smaller screen of walk-ins, most of whom are currently sitting against a windowsill, patient. I'd have to lean in to peer up and I'm not getting in the way of two young men with pans of beef dripping heated to 179 degrees Celsius.

'You've got two light bites, a cod, and a lightly battered regular.' John has prepared another tray of fish to fry. The friers get a batch that is made up depending on the orders. I study to try to understand how John makes the choice of what should go together, but it is as baffling at the end as at the beginning. Instead, I ask profound questions such as, 'Don't the screens stink of fish?' as I watch John swipe away an order on screen with the gloved hands he's just used to pick up dead cod. He answers without answering. 'They're wiped down every night.'

A light bite is a smaller cut of fish. On the website, there are six options for the fussy and particular. Fish can be lightly battered, lightly cooked, well done, fried in rice flour, poached in milk, or poached in water. Miller's offers a modern and broad menu but there are limits governed by science. When an order comes through for a lightly battered fish poached in milk, I hear David swear because this is an impossible request to fulfil. 'They've just clicked on too many options.'

The small restaurant is slowly filling up. I watch Scott preparing two plates and he is counting out chips. What's going on? 'The lady wants five or six chips.' Scott and Ellie explain. It is a generational thing. Older people have smaller appetites. It happens frequently enough that it is now a catchphrase among the staff, who are entitled to eat after their shift but when they don't want much, they also ask for 'six chips please'.

There is such an effort to make fish and chips seem healthy. Baked, poached, lighter. Six chips. David says the light bite (100 grams) and chips is only 1,000 calories. He does not say how many a jumbo haddock and large chips are and there is no calorie information on the Miller's website or app. The health or otherwise of fish and chips has been discussed since the donkey and the barrow.

In 1927, an unnamed reporter for the *Birmingham Gazette* visited a fish and chip shop. His article was subheaded 'Popular dinner of the poor', and he wrote in a way to pull every string of every heart. Here in the unnamed shop in an unnamed part of town was a little lad called Georgie who, 'with amazing patience, born of a long and bitter experience, . . . leaned his small body against the tall counter and waited. He was the merest atom in that small crowd of impatient adults, and yet in some extraordinary way he stuck to the wall of the counter and could not be shifted.' Georgie was dressed in rags and a greasy cap; his boots 'defied description'. He waited his turn, occasionally falling asleep in the warmth, crowded by adults who were 'simply larger editions of Georgie . . . a sprinkling of factory girls, a few tired-looking mothers, some children and a number of men from the great industrial hives of this city of Birmingham'. Georgie's request was modest: four pennorth o' chips and two pieces o' fish. The reporter followed Georgie home and found that this meal had to feed four people. The father got one piece of fish; the other three shared the other. 'That is all there is and if they are still hungry, there is bread and margarine to fill up with.' The reporter was impressed because this was 'a cheap dinner, ready cooked, and hot'. He wondered why fish and chips are still 'a great stand-by for red-nosed comedians' when, where 'there is poverty of a most appalling character . . . they fulfil a very real need'.[33]

Even so, for decades, medical men persisted in thinking fish and chips unhealthy: some linked it to typhoid fever. Medical Officers wrote regular reports on fried fish shops. One Officer in Wednesfield, Staffordshire, discovered a new syndrome which he called fish and chip poisoning. This was 'due to the oil in which they are fried. The main symptoms . . . are vomiting, diarrhoea, prostration, urticaria, and a very itchy and unsightly rash.'[34]

This antagonism was partly due to snobbery: Fish and chips was a poor person's food: fast fish food, like fish fingers, and equally scorned. People worried about the smell, which was considerable because the frying medium was stinky cottonseed oil; the litter; and the wisdom of wrapping food in old newspapers when you didn't know where the newspaper had been. Just before the war, the *Hull Daily Mail* asked, 'Is it an offensive trade?' and 'Is it inimical to public health?'[35]

Sending children out for fish and chips, according to Blaenavon's Medical Officer in 1928, was due to 'lazy mothers' and the direct cause of the town's high mortality rate (rather than bad housing, poor sanitation, and dirty water).[36] This vulgar meal was also protein and fat and carbohydrates and omega-3s. It was vitamins B6 and B12; iron, calcium, and phosphorus; and iodine, fluorine, zinc, and fibre. And it was filling food for empty stomachs. It was a societal good. A national sea fisheries report concluded in 1914 that 'comparatively little fresh fish is consumed by persons of small means' outside fried fish shops.[37] In 1968, it was still fine for Philip Howard to write in *The Times* of the 'plebeian ambiance' of fish and chips at an NFFF awards ceremony in a fancy London hotel (where he also lamented that fish and chips would have been better than the pretentious prawn canapés).[38]

On Miller's website, there are attempts at a soothing supply chain. The cod and haddock, I read, is 'sustainably line-caught by one vessel deep in the Barents Sea'. I don't like to think of a vessel deep in a sea, but I'm also cynical about the 'line-caught'. People hear line-caught, a fisheries academic tells me, and they imagine a man with a rod and line. It sounds friendly. It is a valuable phrase: British consumers were found to be willing to pay 22 per cent more for line-caught cod and haddock. But line-caught is not a man with a rod and line. It is men with lines 100 miles long that carry thousands of hooks.

In the 20 minutes you are waiting for Lewis to fry your fish, you might head to the Marine Conservation Society's 'Good Fish Guide', and you would read that Northeast Arctic cod is a good choice. Then you would click on more information and find that even with its 'good choice rating', the Northeast Arctic cod fishery 'is not in an overfished state, but fishing pressure is slightly above sustainable levels'. Norwegian cod caught within Norway's 12-mile limit recently lost its Marine Stewardship Council (MSC) status. Cod migrating from the Barents Sea towards Norway is fair game: the less roaming coastal cod, visually indistinguishable from its highly migratory fellows, is not. Because fishermen were catching the coastal variety by mistake, cod got demoted.

Fast Fish Food

To understand where your fish came from, how it was caught, and whether it is sustainable, terrible, or criminal takes effort. In supermarkets now I head for the fish fridge and spend several minutes reading the story told by fish packaging. My phone fills with photos of labels of fillets and cans and prawns and scampi. I learn that there will always be an attempt to give a sense of provenance. It is mostly trickery. Or as Steve Trent of the Environmental Justice Foundation says, 'It's all rubbish.'

I pick up a packet of chilled skinless and boneless cod loins and read on the back of the packet that it was 'packed in the UK using cod caught in the North-East Atlantic using Trawls, Hooks and Lines, Seines, Gillnets and similar nets'. Even if you speak fish, this will seem extraordinarily broad, like saying that a potato was dug up from somewhere between Brussels and Bulgaria using a hand, a spade, or an Abrams tank. As information goes, it is so generous as to be useless. The front of the packet is more help: the catch method is listed as trawls. Pelagic? Demersal (better known as bottom trawling)? The packet does not clarify, although these two types of fisheries have enormously differing impacts on the ocean. Instead, maybe I'll chose a cod from the Icelandic fishery, one that is caught by gillnet. I check the 'Good Fish Guide', downloaded onto my phone so that I can stand in front of the fish fridge for hours checking everything I need to. 'The Marine Stewardship Council (MSC) indicates that harbour porpoise and northern fulmar by-catch are the main concerns for this fishery . . . 1,500–2,000 harbour porpoises are caught annually in gillnets around Iceland.' The harbour porpoise population is of 'least concern' on the International Union for Conservation of Nature (IUCN) Red List, so those dead porpoises are meant to be tolerable. Also, the MSC is heavily criticized for allowing – for example – fisheries with huge amounts of by-catch, and for having loopholes that allowed shark finning until 2022, when the organization issued an updated guideline that requires fisheries that retain sharks to follow a 'fins naturally attached' policy. Nor does the world's most popular fisheries certification address CO_2 emissions or labour conditions or what happens to the fish that are fished. 'The MSC,' wrote Jennifer Jacquet and Daniel Pauly in a paper called 'Reimagining Sustainable Fisheries', 'have certified Norwegian fishing vessels that catch krill in Antarctica that are fed to farmed salmon, which, in turn, are destined for luxury markets. If French companies were killing penguins to fatten the geese used in foie gras, the public would balk at the idea of "sustainably caught penguins", regardless of any outcome suggested by a penguin stock assessment.'[39]

None of that is mentioned in the 'Good Fish Guide'. Navigating the guide is already complicated and I'm not even buying the fish. Who has the patience to spend 10 minutes on an app to choose an environmentally safe dinner? Maybe the Americans do it better. I choose a packet of salmon on Walmart's online site that is labelled as GGN-certified. I look up GGN (it's a certification programme) and input the 13-digit number. There is no QR code. The number takes me to a friendly page belonging to Skagerak Processing in Denmark. Then if I want to know where the fish came from, I have to scroll down to its suppliers, and it could be one of three. Each supplier has a GGN number but none correspond to the number on the label. Maybe a domestic fish will be more straightforward? A packet of 'US farm-raised catfish fillets' has a barcode but no other details beyond 'proudly farmed in the USA'. The allergen statement reads: 'contains catfish'.

Perhaps the best attempt at explaining the complicated and turbid seafood supply chain was delivered by a young man in Bizzie Lizzie's fish and chip shop in Skipton, Yorkshire. When I asked him where the fish was from, he looked at me impassively. 'The sea?'

Every so often, the trade association Seafish issues a 'Fish and Chips' report. The last was not cheery. Fish and chips at the end of 2022 was still Britain's favourite takeaway in units sold, yet there had been 1.2 million fewer servings of fish and chips.[40] For those fulminating reporters and medical officers who for more than a century derided the poor person's meal: 69 per cent of fish and chips are now bought by people from the A, B, and C1 consumer, occupational-based categories used by the Office of National Statistics that include professional and managerial occupations (A and B) and supervisory, administrative and also professional occupations (C1).[41] At £8.99 ($11) on average, fish and chips has lost its 'persons of small means' tinge, although when I find a fish and chip menu that under 'sides' lists chewing gum for 50 pence (60 cents), I think it must survive at least in the working-class suburb of Middleton, Leeds.

When 1,600 Britons were asked what their favourite takeaway was, one in four chose Chinese, then Indian, and only then fish and chips.[42] The demographic breakdown showed that fish and chips is an older person's choice, with the majority eaten by over-50s. Seafish has recommendations to fix this: by building 'on a unique foundation of enjoyment, highlighting health and quality credentials (better living) whilst educating about the different types of species available (choice and convenience)'.

The NFFF website proclaims that the UK's fish and chip shops serve the same volume of fish and chips as in the postwar years, have revenues of £1.2 billion ($1.5 billion) a year, and consume a third of all whitefish sold in the country. 'We definitely punch above our weight,' says Andrew Crook of the NFFF. 'I always say that we're the second most British thing after the Queen. And now she's gone . . .' Despite David Miller's healthy turnover, the industry is wounded. Hikes in gas and fuel prices have pushed some shops to close. Everything now costs more – fat, fuel, potatoes, fish – but shops can't afford to charge less. The vinegar manufacturer Sarson's declared that half of all fish and chip shops could close by 2025[43] and the NFFF worried that the industry was facing 'an extinction event'.[44]

After my shift, David sends me screenshots of his WhatsApp chat with his staff where they record sales and wastage. He sends me numbers. During the Friday shift I attended they sold 193 haddock, 65.4 cod, 296 portions of chips, 63 jumbo haddock, and 65 light bites with a side. Haddock sells better now, he says. He's not sure why. He will serve what is wanted. Because David Miller, fourth-generation frier, has no intentions of closing. Fast fish food or not, his is a noble trade and no one can tell him otherwise.

First image by salvage team of Scandies Rose

4

We're Rolling Over

*Dear Lord, be good to me, the sea is so large
and my boat is so small.*

BRETON FISHERMAN'S PRAYER

The order of the watch went like this: Seth, Brock, Art, David, Jon, Dean, then Gary. The crew did an hour's watch each; Gary Cobban was the skipper and did six hours at a time because that was what skippers should do. David Cobban was Gary's son, Arthur Ganacias was the engineer, Brock Rainey had fished with Gary for 15 years or so. Seth Rousseau-Gano came from Washington State, as did Art. Dean Gribble was new but only to the boat; he had been fishing since he was 11 and his friend Jon Lawler had asked him to join. The boat was the *Scandies Rose*, and it set out to sea from Kodiak, Alaska on 30 December 2019. That evening the weather was as bad as December could throw. The report, said Jon Lawler in later testimony, 'was enough of a shitty forecast to . . . I didn't think we were going to leave that night'.[1] Fierce, frigid winds were blowing at 8.35 p.m.[2] Some boats stayed put but *Scandies Rose* still set out. 'We knew the weather was going to be bad,' said Dean Gribble, deckhand, 'but the boat's a battleship, we go through the weather.'[3]

The boat was loaded with fuel and seven men. It carried 198 cod/crab pots and 7,000 kg of bait and was headed northward towards False Pass and the Bering Sea. She was trim, said Dean, and a good boat. Gary was a good captain. *Scandies Rose* was the sister boat to the *Patricia Lee*, which Dean's dad skippered. Dean had the wariness of a new crew member, and after due consideration he had no concerns about crew or boat.[4]

The last jobs before departure were the usual: get the food, water, and fuel on; stack the pots properly. 'The pots were stacked great,' said Dean. Every row had a chain on it. 'I thought it was a little overboard, but now that I look back on it, yeah, it was probably good. And it was tied down a lot. There was ties everywhere. They were tight.' Dean didn't count them, how could he? Gary told him there were 198 on board. That is a heavy load but not unusual. Each pot measured over 2 by 2 metres. 'Big, heavy fucking pots.' (Dean uses the same curse word as an adverb, adjective, noun, preposition, and any other grammatical element. They weren't redacted in his testimony, so I'm keeping them too.)

These pots were giants compared to the crab pots I knew from fishing off Yorkshire. So the stack was huge and Dean learned that fact quickly because he had to climb all over it. It was four or five pots high and with all his checking and checking, his legs and arms were burning. It had to be done. Crab pots stacked badly could affect stability, and lack of stability can sink a boat. In the 1990s, the US Coast Guard began doing pre-departure stability checks in Alaskan ports, which reduced fatalities and sinkings by almost 70 per cent.[5] Gary expected conditions to be 'icing', and a crab pot laden with ice can weigh more than 1,000 kg. Any trim or stability would be thrown by 198 times that amount. 'Fuck,' said Dean. 'It seemed fine.'

The captain did a safety drill. This covered where the emergency position-indicating radio beacon (EPIRB) was located, how to make a Mayday call, and where the fire extinguishers were. A crew member demonstrated how to put on an immersion (survival) suit. Immersion or survival suits are waterproof full-body garments with a hood and integral gloves and boots. They are bulky and hard to get on but far more likely to save your life in cold water than a lifejacket. The drill was documented, signed by all the crew, and sent to the vessel manager ashore.

Scandies Rose departed. First north-east then north-west towards hard fishing grounds made worse by weather, towards the Bering Sea.

We're Rolling Over

The forecast looked like this: wind 15 knots, becoming SW 25 knots in the afternoon; seas 3 feet, building to 6 feet in the afternoon; freezing spray. There are many people who listen to the BBC's *Shipping Forecast* because they find it soothing. That is because they are not where the weather is. There was nothing soothing about that forecast, and 'shitty' is an inadequate word for its power.

The first day was testing but OK. On 31 December, Dean's hour-long watch finished at 7.15 p.m. He woke Gary to take over and told him the boat had a bit of a list: it was leaning to one side. This had to be commented on but wasn't necessarily bad news. 'It wasn't too crazy. Maybe one degree.' At 7.15 p.m., when Dean handed over to Gary, the wind on the forward and starboard beams was approaching 60–70 knots. On the Beaufort wind force scale, a measure of wind at sea and on land, this wind speed counts as 'Violent Storm' and only 'Hurricane' is stronger. For Violent Storm, the wave height is 11.5–16 metres (about the length of a big truck) and 'visibility is greatly affected'.[6] On *Scandies Rose*, the crew who weren't down below in the accommodation didn't need the Beaufort scale. They could see the weather through the windows, at least for now.

Soon the freezing spray was hitting the starboard side of the boat because of where the wind was coming from. 'We were building ice,' says Dean, 'but again it wasn't anything crazy.' Gary was aware of it, and Dean asked when Gary came to take the watch whether they should go outside to try to chip the ice off. This would be standard practice if the weather allowed. Gary, a cautious captain, said they would wait to get to shelter, and Dean was relieved. 'It was shitty out and I didn't have any qualms.' He went below, had something to eat, and then headed to his 'rack', as fishermen call their bunks, and began to watch a movie.

In the United States, if you want to die on the job, you should become a logger or a commercial fisherman. These two jobs occupy the top spot on the highest risk careers; one year it's one, the next it's the other. The National Institute for Occupational Safety and Health calculates that fishing is 40 times more dangerous than any other job.[7] England's

Marine Accident Investigation Branch (MAIB) puts the rate of fatalities in the fishing industry at 100 per cent higher than in the UK general workforce, with 85 per cent of those involving people ending up in the water.[8] Some of the headings in the MAIB's latest safety digest: 'It was all very fast'; 'It's not you, it's them'; 'I wouldn't stop there if I were you'.[9]

But for a country that is thought to have the best-regulated fisheries in the world, the United States is an odd fish when it comes to mandatory safety requirements. It was only in 1988 that it became mandatory for US boats to carry safety equipment such as life rafts and an EPIRB, a locator beacon that should detach on sinking then float.[10] UK fishermen must complete four safety courses as a condition of their licence.[11] US fishermen do not need a licence unless their vessel is over 200 US tons. 'That's a tiny percentage of the fishing fleet,' says Jerry Dzugan, a former fisherman who provides safety courses in Alaska. 'Vessels this size are built to or reconfigured to be 199 tons to avoid having licensed crew.' The only mandatory safety training is an 8-hour course instructing how to use survival gear and conduct an emergency drill, but that is only for boats that carry more than 16 crew or operate in certain latitudes. After that, they have monthly drills that can be carried out on board or ashore; they don't have to be done by a crew member and enforcement and oversight varies widely depending on geography.

Above 32 degrees latitude in the Atlantic Ocean, commercial fishing vessels must carry immersion suits. Below that latitude, they must carry lifejackets, also known as personal flotation devices (PFDs). In other waters, the dividing line is 35 degrees.[12] The PFDs have to be on the boat. Fishermen do not have to wear them.

Dzugan has worked in fishing safety for 40 years. He has also fished: salmon seining, salmon trolling (long-lining), then Pacific halibut trolling. We talk over Zoom: he is in Sitka, I am in Leeds. He wins. In 2010, he tells me, the US Congress passed a law saying that fishermen are required to have training in stability, navigation, rules of the road, and a couple of other things. All sensible things when you work at sea. 'But

We're Rolling Over

the Coast Guard has yet to enforce it. Fourteen years now.'

Between 2000 and 2020, there were 805 deaths, 164 missing people, and 2,122 people injured in commercial vessel accidents in the United States.[13] Of the 210 people who fell overboard and drowned, not one was wearing a PFD.[14] That's bewildering. Why would you not wear something that can save your life? 'It's a cultural thing,' says Jerry. 'I remember when I first was fishing in the 1980s, I would go on boats and I would ask if they had an immersion suit, and one guy threw me off his boat for even asking the question. Another example was somebody I asked . . . where the life raft was, and he said we don't talk about that on this boat. It's like, if you talk about it, bad things will happen.'

I know about fishers' superstitions. Never say rabbit, but call it underground mutton (this is something to do with the Devil appearing as a bunny). Don't whistle on board. Never wave a fisherman off to sea because a wave will bring a wave. Jerry has some new ones. 'Bananas,' he says. 'Though it depends on the boat, sometimes it's apples.' There are a few theories about the banana aversion: because they ripen and ethylene puts off the fish; because banana boats often sank; because fishermen slipped on the peels. Anyway, you would only know once you got on board and your fruit got lobbed into the sea by an irate captain.

These superstitions are not about ignorance but hope. 'Fishermen work and live in an environment that they don't have control of,' says Jerry. 'They don't have control of the fish, where they are, or how to catch them. You don't have control of the weather. You don't even have control of your boat in a lot of senses in that you try to maintain your boat, but unforeseen things happen: electric fires, sinking, flooding. You don't expect it.' Superstition spins a web around the uncontrollable. Superstition is containment. Jerry knows fishermen with graduate degrees in marine biology who won't leave harbour on a Friday. One survival suit manufacturer lost business when he put his survival suits in a black bag. You don't bring a black bag on a boat. 'It's a body bag.'

All hands on deck, when it comes to improving your chances at sea.

That would mean wearing a lifejacket. Even if you fish in waters where you carry an immersion suit, which you would never wear unless about to be immersed, why would you not wear a PFD on deck? Jerry counts the ways. They're bulky. You can't work in them. They have straps and webbing, and straps are anathema on a working deck – they can snag and they can get you killed. That's the practical stuff. There is also the machismo. 'And a lot of it is from, "I've been fishing for 50 years, nothing bad has ever happened to me – why should I worry?"'

Yet every fisherman that Jerry knows, and every fisherman elsewhere too, knows someone who has died at sea or had a near miss. 'You realize you know so many people who have lost boats. You have lost so many friends. We all know people who have been lost.' Still this does not translate into wearing a PFD. 'I know one fisherman who has been overboard twice, and his son saved him, and he still doesn't wear a PFD.'

The RNLI publishes a leaflet called 'Lifejackets: useless unless worn'.[15] Fishermen, I read in one academic study, have a tendency to 'trivialize or totally deny the dangers associated with their occupation. When they admit it, fishermen have a tendency to claim that danger affects other fishermen but not them, because they are careful.' Even when they carry the necessary safety equipment, 'the compliance is grudging and superficial at best. Some fishermen reportedly keep EPIRBs in a drawer in the cabin to protect them from theft but forget to place them back in the bracket when at sea.' One fishermen told the researchers that his brother died in 30 fathoms of water 'with two EPIRBs stored in the cabin'.[16]

Back in Alaska, the weather was so atrocious that a Coast Guard rescue pilot drove to his station with his head out of the truck window so that he could see. On *Scandies Rose*, Gary Cobban was on the phone. He phoned Jeri Lynn, a woman he had been seeing, to wish her a Happy New Year. She said later he sounded fine. They talked for 15 minutes. He spoke about needing to 'tuck in' somewhere safe. He didn't sound stressed or worried. 'He told me the boat was icing and it had a list to it, but he didn't sound alarmed. He didn't sound scared. The boat ices. The boat ices every winter. It's just something they deal with.'[17]

Gary's sister Gerry described her brother as a man who had fishing in his blood. Gary's dad began fishing when he was eight or nine years old. Gary's mum fished. Grandparents fished, parents fished, children fished. By age 20, Gary was already a captain. He had done a season 'on the back deck' the year that *Deadliest Catch* started. Other than that, said Gerry, he was in the wheelhouse. He had a blip in 2007–2008, 'and he was like, I just don't know what else I would ever do in my life. What am I qualified to do?' Gerry listed some things. 'You know how to weld and you know how to do mechanics and you know how to paint and you know how to organize things . . . there is life outside of being a skipper.'

Gary did not leave. 'Although my dad still ran a boat at 76 years old,' said Gerry, 'it wasn't crab fishing in the Bering Sea, you know?'[18] *Deadliest Catch* chose to focus on the Bering Sea crab fishery because at the time it was the deadliest. That is no longer true. In Alaska recreational boating is more deadly. When Jerry and I speak on Zoom, he tells me that four boats have gone down in the last 13 months, all from Sitka, population 8,000.

When Gerry and her husband went to dinner on the evening her brother left for the sea, her husband looked at the weather. 'I bet your brother's just getting his ass kicked out there.' Yes, said Gerry, because the weather was what the weather was. 'But he's a big boat and he knows what he's doing and he's going to be OK.'

Gary could be exacting at work and outside it. He talked often of 'the effing crew' not getting things right. He was hard on his son David, harder than on the other crew. Gerry called Gary 'Junior'. 'I remember saying one time, just a brief conversation with Junior, in all my frustration, saying, hey dude, slack off on David. He's not drinking to excess, he's not doing drugs, he's never been in jail. So what if he doesn't have a car?' Still, Gary was a good captain on a good boat. He did the monthly drills as he was supposed to, so the crew knew where the immersion suits were, where the EPIRB was, how the life rafts worked. He was rigorous, said Dean. Thorough. A good person to be in charge.

In the *Scandies Rose* wheelhouse, on the last night of 2019, only two windows were heated and still functional as windows. The rest were obscured by ice. Peering and straining to see, the captain would have seen that pots were iced, but he could not have seen how much. Only going out on deck would have revealed the thickness of the ice on them. Even had that been possible, the pots had not been stacked with access alleyways built into the stack, as the Coast Guard advises. The crew would have had to clamber over the stack to monitor it effectively. This was unthinkable in this violent, turbulent sea.

'Some people,' I read in a piece about fishing safety, 'see fishing as a relaxing activity, where the biggest danger is wet feet.'[19] The most accepted figure for how many fishermen die each year came initially from the International Labour Organization and was 24,000. The FAO upped that to 32,000.[20] In 2022, the FISH Safety Foundation, after two years of study, came up with 100,000 deaths a year.[21] Debate still moves around these figures although it need not. Too many fishermen die when often they don't need to.

What is known is that most deaths happen to people fishing in small boats. What is also known is that fishing has been made safer, but it will never be safe because fishermen deal with weather and slippery surfaces and heavy machinery and luck. Danger can never be legislated into nothing, and fishing safety legislation is decades behind laws governing shipping. The most important maritime safety convention, the International Convention for the Safety of Life at Sea, does not apply to fishing vessels except for some of its navigational safety sections.[22] The Cape Town Agreement, legislation aimed at fishermen that would apply to small boats, was signed in 2012 but has yet to be ratified because of insufficient signatories.[23] Cor Blonk, a Dutchman and former policeman who now campaigns to improve fish safety legislation through the FISH Platform, thinks this is because the major international organizations are biased towards shipping. 'What you see is that representatives in administration and at UN bodies like International Maritime Organization almost always have a background of merchant shipping.' They don't understand fishing, so they don't consider it. This is numerically nonsensical. 'People are not aware that about 1.8 million seafarers work on commercial vessels. And the FAO estimates that there are 38 million fishers around the world. So that means that every 20 out of 21 people working at sea are fishing.'

I was drawn to the *Scandies Rose* because the boat had my name. There was also the *Emmy Rose*, a fishing vessel that sank off Portland Maine a year later: four crew dead. I would have liked to list other vessels here but the United States has no publicly accessible database of fishing casualties. Experts know that sinkings are diminishing and crew overboards are increasing. In one official fisherman's safety guide, a section is entitled 'What can go wrong'. The answer is 'hazard'. And what is a hazard? 'Almost anything.'[24]

On a stable deck of a boat on a calm sea in a calm situation, it can take 60 seconds to put on a survival suit. That's what Jerry Dzugan teaches people to do at the training courses he runs at the NGO Alaska Marine Safety Education Association (AMSEA). First you have to pull it out of its bag and hope you've got the right size. *Scandies Rose* carried four 'adult-size', three 'jumbo', and one 'intermediate'. Then hope you have one that fits. Immersion suits work like dry suits, not wetsuits: water is not meant to ingress. A suit that is too ample means the hood could float up and let in water. In one sinking near the Aleutian Islands, wrote *Fishermen's News*, 'everyone survived in survival suits except the captain, who could not get the zipper up due to his large midriff. He entered the water with the zipper open and perished due to cold water.'[25]

Then, lay the suit out flat. Take the two plastic bags in the hood to put over your boots so they slip in to the suit more easily. Sit down to step into it, slide your boots in, then stand up and pull the top half over you. In an instructional Coast Guard video, it looks like less trouble than a wetsuit, but not by much.[26] You don't shimmy into an immersion suit. Now insert your fingers into the cartoonish attached three-fingered gloves, and now that you have less dexterity, you need to make sure the suit is on right, pull up the huge zipper, then get the hood up, and Velcro it across your face. (In AMSEA's video, you put the hood up much sooner, a tip taken from interviewing survivors.) The Coast Guard video lasts for just over two minutes, on a boat that is not pitching and listing in a howling gale with icy sea spray, in a wheelhouse crowded with your terrified crew members, all convinced they will soon die. The boat in the Coast Guard video is definitely not what have been called 'wet rolling vessels'.[27]

Jerry acknowledges that the training is imperfect. 'We're doing it on a stable surface. And that doesn't count the adrenaline rush when you're abandoning ship quickly. So yeah, it's not the real world.' A shaky video online shows an event from a Fisherman of the Year competition (also not the real world).[28] Onstage, two fishermen and a fisherwoman are competing to get into a survival suit as quickly as they can. The music is 'Danger Zone' by Kenny Loggins. Jerry would be pleased that the winner is a man from Sitka, who gets the suit on in 37 seconds. The presenter asks him for tips and the Sitka man is efficient with his answer. 'Legs in, weak arm, hood, strong arm, zipper, and flap.'

A translation: you get your weak arm in the sleeve and attached mitt first so that you can better manipulate the rest of your body into the suit with your better arm. Online there are other tips for real-life situations: on a rolling boat, start on your knees and lie down for the rest of it. If you have stability, you lie first then stand up. An immersion suit is not perfect, but it multiplies your chances of surviving by six times.[29]

In his rack aboard the *Scandies Rose*, Dean was still watching a film. Then the boat leaned hard. 'I looked up. Jon got up, ran upstairs, yelled down, "Dean, boat's sinking."' Dean got his trousers on and ran up to the wheelhouse. He thinks everyone was there. Dean threw out the survival suits to everyone from their locker. He doesn't mention the different sizes. Everyone was just trying to get them on. He remembers that Gary looked 'freaked out'. 'We all were. This is horrifying – it's the fucking worst type of weather you can even fucking think of for this shit to happen in, and it's fucking happening. It's blowing fucking 30, 40 miles an hour. And there's fucking ice all over the windows, so it's fucking cold. And now I've got to go swimming in this shit.'

It was so hard to get the suits on because the vessel was now listing by 20 degrees. Dean got his legs in his suit by jumping into a chair in the wheelhouse and bracing himself against the armrest. He slipped, got back up, got his suit on, zipped it up. He went outside, helped Jon with his zip. They made a plan. They shouted it to be heard over the wind. 'Fucking stay together. We fucking live. We're not fucking dying. We stay together.' Later Dean said, 'I knew in my head we were dying.'

He looked back at the wheelhouse and saw David Cobban, Gary's boy, at the door. He had his suit on, but he wasn't moving. Dean screamed at him. 'Get out! Get out! We need to get out! The boat's fucking sinking!' Everybody was in shock. 'You can't fault them for that. It's a fucking horrifying fucking thing that was happening.'

Everyone was screaming. Dean was screaming at them to get out; they were screaming for help. From whom? Gary Cobban had sent a Mayday call at 21:55. He repeated 'Mayday' several times and gave the vessel's name and position: 56° 29' N 157° 01' W. On the recording he sounds calm when he says, 'We're rolling over.'[30]

Fall into cold water, and you will instantly gasp at the cold and breathe in water. Your body will immediately transfer blood away from your extremities to keep your core warm. This is called the cold shock response, and it means that the first few minutes are critical. They are less critical if you are wearing something to keep your head buoyant and out of the water. People who can hold their breath for a minute in air can only hold it for 10 seconds in cold water. 'Consequently,' wrote Frank Golden and Michael Tipton in *Essentials of Sea Survival*, 'in choppy or turbulent water where small waves may intermittently submerge the head (and airway), a person has a significant chance of aspirating water during the first few minutes, until he or she can control respiration.'[31] It takes less than half a litre of water to drown you, but you only need 3–4.5 kg of buoyancy to keep you afloat. On the lobster boat *Beverly* off Massachusetts in 2019, a crewman was working on a line when it snapped. Captain Mike Bartlett turned round to see him overboard and 'his legs already over his head'. The crewman was wearing flotation bibs, made by a company called Stormline; they look like everyday fishing oilskins, but they float, because they contain 5 kg of buoyancy. The *Beverly* was turned round and reached the overboard within a minute. By that time, 'he had no use of his hands and legs. He wouldn't have been afloat if it wasn't for those [bibs].'[32]

Julie Sorensen is director of the Northeast Center for Occupational Safety (NEC), which works with fishermen, farmers, and foresters. In 2016, it launched Lifejackets for Lobstermen, a project to improve lifejacket use (fishermen use the word 'lifejacket' more than 'PFD'). Rates of use were as poor as in any other fishery. One lobsterman interviewed had been fishing for 40 years and never worn one. The project involved social marketing and outreach. The questionnaires can be summarized as follows:

1. Tell us everything you hate about lifejackets.
2. Tell us everything you would change about them.

Vans were equipped with a variety of lifejackets and driven to ports, where the lifejackets were offered at half price. In a six-month period, 1,087 were sold (including to the fisherman who had not worn one for 40 years; he bought some for himself, his wife, and his crew). The designs were carefully chosen. Not all were Coast Guard-approved, but a non-approved lifejacket is better than no lifejacket. There were fishermen's bibs that incorporated flotation and looked nothing like a lifejacket and everything like everyday fishing workwear. There were regular-looking lifejackets, waist packs, and even a T-shirt. Anything that had a pull tag or strap hanging down was disliked. Anything that laid flat and could deal with a 12-hour day of constant movement was acceptable. The bibs sold most. In the summary report of the Lifejackets for Lobstermen project, one chapter title is 'The Best Lifejacket Is

We're Rolling Over

the One You Wear'. A headline on the NEC's website reads 'SURVIVAL OF THE FITTED'.

> Almost lost my old man overboard last week. He went in, and I wasn't sure we were gonna get him back. He had his muck boots on. They kept floating up and his head kept going under. It was scary. I gaffed him by the hood to get him back in.
>
> PETE SEIDERS, SON, TOLD AT THE
VANS' STOP IN SOUTH BRISTOL, MAINE

For Sorensen, the growing problem is not vessel disasters (which are decreasing), but how to get a crew member who has fallen into the sea out of it again. That's if the overboards are noticed in the first place: in the case of the 204 commercial fishermen in the US who died overboard between 2000 and 2016, 59 per cent of the falls were not witnessed, and 108 bodies were never found.[33]

'I've had fishermen tell me they've had crew members that have gone overboard and they've desperately tried to get them back on board,' says Sorensen. 'But when your boat has lost meaningful movement and you're trying to search for somebody in the water and you've got very little time because the water is cold, lots of things have to work really well.' She knows fishermen who have had to let go of their buddy. 'Then they live with that for the rest of their lives and it's horrific.'

She worries even more about solo fishermen. There are more of them because fishermen are under such huge financial pressure – mortgages, fuel, boat loans – that they dispense with crew. The trouble with that is that if you go overboard, how do you get back on? Some fishermen have kill switches so that their engine stops if they leave the boat. In the UK, official advice to solo fishermen is to attach tyres to the side of the boat, the better to clamber on with. Or a ladder. 'If your extremities are frozen, how do you use a ladder?' says Sorensen. She would prefer to see boats fitted with lifts. If you can't afford crew, how can you afford a lift?

Sorensen thinks US safety regulation for commercial fishing in the US is 'a hot mess', and quotes Winston Churchill to back herself up. 'You can always depend on the Americans to do the right thing once every other option has been exhausted.'[34]

Among the chaos and fear, Dean was trying to get the light on his suit to work but his hands were in 'these stupid gloves, these Gumby gloves'. He couldn't manipulate the light out of the pocket it was in. He was desperately trying to find some buoys or anything they could hold on to in the water. He was trying to throw a line to the wheelhouse so that the men inside it could climb back up. I read this and try hard to picture the topsy-turvy physical reality of a boat about to sink. Climb back up where? By then Dean thought the boat was listing by 45 degrees. The power cut out.

Dean was still yelling at David to get out. 'He was just sitting there in kind of shock. He probably didn't want to leave his dad. I've fished with my dad. I don't know what I'd do in that situation either. I don't know if he made it out. I hope maybe he was washed out. But the last time I saw him, he was just sitting at the door and I was screaming for him.' The boat was so skewed now that Dean and Jon could walk on the side of the wheelhouse. Dean was shocked at the speed of the disaster. He thought the boat would lean and list for a couple of hours. In fact, 'between sleeping and swimming was about ten minutes'.

A wave washed Dean and Jon into the ocean. They were separated. 'I'm getting tossed and turned. Fucking water all down my stupid suit. And I'm going to die. This is how I'm going to fucking die, New Year's Day.' He had the ability still to wonder about the details. 'Do I just suck in water? Do I just wait to freeze to death?'

Here I think of a tale Jerry Dzugan told me, of a fisherman's supply shop he visited in New Bedford, Massachusetts. He saw that it stocked steel-toed boots and asked the shopkeeper whether fishermen bought them, because they made no sense to Jerry. Waders, yes, not steel toes. 'Yes,' said the shopkeeper. 'Because they don't want to suffer if they fall overboard.' I see survival videos showing how to turn your trousers into a makeshift lifejacket (you tie the legs together, whack them over your head so they fill with water and keep them wet). Some fishermen don't want that. They don't want to float if they can't be rescued. They don't want to float if it's just a slower way to die.

In the water, Dean was screaming for Jon. He was getting tumbled by waves but in between the tumbling, he could see *Scandies Rose*. He couldn't quite believe it and in his testimony his cursing gets even more frequent to convey his disbelief. 'The fucking bow fucking vertical in the sky. It just goes straight down like the fucking *Titanic*. I'm just sitting there in shock, like, what the fuck? And whoosh. Gone.' He didn't know this, but they were 2.5 miles from Sutwik Island, where Gary had been heading to seek shelter.

Dean started to surf the waves as best he could. If he was going to die, he may as well have some fun with it. So there he was, riding the waves, pretty comfortable in his suit, when he saw 'a beautiful fucking survival raft floating right toward me'. It was 15 metres away and Dean would have swum all night to get to it. He rode a wave that took him right to the raft, and he tried to get in. 'But fuck, you know, I was tired.' He sat there for a minute gathering his strength, then pulled himself in. He was screaming for Jon and then he heard Jon scream back. 'Now we're in the raft. At least our bodies will get found or whatever, depending on how long.' They thought they would be found because they thought the EPIRB would have been triggered. 'Apparently it fucking didn't. That's awesome.'

A life raft is equipped with some food, water, and safety equipment such as flares and a flashlight. They shot off a flare and it hit the sea. They tried another one. Too soon, in hindsight; they had only been in the raft an hour or so. 'But we thought we were dying so we just freaked out.' Dean thinks there should have been more flares. 'All the survival equipment in the raft, it's fucking worthless. The raft itself? Beautiful piece of equipment. The things in it were just junk.' For a start, everything was wrapped up and tied. How can you untie anything when you are trapped in a survival suit with its Gumby gloves?

Are you supposed to take off the suit that is keeping you alive?

The raft had four feet of water in it. That didn't mean they would sink, but it meant they were cold, and then they were iced. All the supplies were rolling around in water they had to duck into if they wanted anything. Dean sounds furious in his testimony. Why don't they make flares that float? 'Bullshit.' He had 80 packets of fresh water, but you can't fire fresh water into the sky to be rescued. The lights in the raft went out within 10 minutes. 'Why aren't there LED lights all over the fucking thing, just glowing? Why doesn't the suit have a headlamp so you can see shit? It's just very lacking.'

And the EPIRB that should have transmitted their location: why not have two, one starboard and one to port? The single EPIRB on *Scandies Rose* had been outside the wheelhouse on the starboard side. It should have gone off, and the bracket was later found to be empty.

Dean and Jon didn't know about that. They expected to be found quickly. 'We're sitting there waiting for a ride. Kodiak is right there. Boom. In the Uber going back to Kodiak. It didn't work out that way.'

Vessel disasters get headlines and attention and expensive multiday hearings. Actually more American fishermen die from going overboard than in vessel sinkings. 'They are one-by-ones,' says Jennifer Lincoln, an injury epidemiologist and the associate director for the Agriculture, Forestry, and Fishing Safety and Health Department at the National Institute for Occupational Safety and Health. One-by-ones are not noticed. But they are also the reasons for the safety campaigns, for the constant and valiant attempts to persuade fishermen to wear lifejackets when they work, to figure out how to get someone overboard back on the vessel, to carry ladders or gaffers that someone can grab on to. In fact, more crew fall overboard in the shrimping fleet in the Gulf of Mexico than in Alaska. That's where the Discovery Channel should be filming. *Deadliest Shrimper.*

In 2000, Father Sinclair Oubre, a former merchant seafarer who is now a chaplain for Stella Maris, co-founded the Port Arthur Area

Shrimpers Association. He talks to me from the deck of a training ship anchored in Fort Lauderdale. His profile picture shows him with a motorbike, and I am reminded of Farv, a London priest who set up a gang of young motorbikers to transport blood around the city. Maybe having a motorbike makes you a campaigning priest? Back in 1997, someone at the Port Arthur marine safety division got in touch. 'They were running into a problem with the shrimping boats from running into ships or not knowing the rules of the road.' There are hundreds of shrimpers in the Gulf of Mexico, but there are also barges and commercial ships. Once, a tugboat and a shrimper were heading for collision. 'So the shrimp boat's coming down the channel, and the tugboat gives a toot. The shrimp boat moves to the other side, the tugboat toots and the shrimp boat moves again. The tugboat was giving the proper signals, but the shrimp boat just thought the horn meant he was in the way. So he kept moving backward and forward until they finally ran into each other.' No one was hurt, but it was unsettling to realize how little knowledge the shrimpers had of basic maritime safety. The association was set up to improve that. It sounds, over a bad connection from Florida, like a tall order.

If you yearn to be a commercial fisher, despite the costs, the danger, the warming oceans, and everything else, the United States makes it easy. You just buy a boat of up to 99 feet and off you go. No training required. Insurance companies and vessel owners may require safety training but if not, it is a free-for-all. 'As a former merchant mariner,' says Father Sinclair, 'I find that pretty crazy. If I want to run a passenger vessel, I immediately need a six-pack, which is a Coast Guard licence to transport six or fewer people in a boat. A shrimp boat that is 75 feet long needs no licensing.' He still has marine pilots come to the association's meetings to tell the shrimpers such simple things as: 'Don't have your riggers out. We can't move big ships out of the way.'

The shrimpers also need to be told to wear PFDs, because they don't. 'It's too hot. They don't want to.' The shrimp fleet has changed over the last couple of decades. It used to be mostly Cajun and now is mostly Vietnamese, descendants of boat people, and other immigrants. The attitude to safety has not changed with the ethnicity of the shrimpers. Father Sinclair talks of new belt PFDs that are better; you go overboard, you pull it over your head, and you float. 'It's not a safe culture. The thinking is, "We go out there and do the job and we don't bother with this stuff."' He tells me the story of a shrimper who went out to the outrigger – long beams that extend out each side that carry the nets – and fell in. He got back on board, did the same, fell in again. 'The third time, he didn't come back up.' On a merchant ship, Father Sinclair had to do safety drills every 10 days. On a fishing vessel, it's monthly, and there is no requirement to check the log to see if the drill has been done or done correctly. Father Sinclair has seen the merchant fleet transform its safety culture while the fishing industry hasn't. Why not? He has a complex answer: federal regulations that aren't fit; a Coast Guard more concentrated on 'flashy activities' since 9/11; a concentration on marine pollution; and the fact that 'fishermen fight like cats and dogs against any regulation'.

It is standard practice for search and rescue authorities to ask other vessels in the area to assist. Usually, this is done. The weather on the night of 31 December was too atrocious, and when at 23:00 the Coast Guard asked the crabber *Ruff & Reddy* to head to the scene, its skipper refused, as a skipper has a right to do if he believes conditions to be too treacherous. The *Ruff & Reddy* had chosen to seek shelter off Nakchamik Island at 05:00 that morning. By now the boat was icing up and it was too hazardous to leave its sheltered position.

At 23:26, the Coast Guard at Juneau, on the other side of the Gulf of Alaska, took control of the search and rescue, after it was judged too complicated for Kodiak. Four minutes after that, helicopter CG 6038 with a four-man crew was airborne. A plane followed. The helicopter's pilot testified that because of the weather, 'this was my most challenging flight of my career just getting out there'.[35] There was half a mile of zero visibility and severe turbulence. Then, as they approached the *Scandies Rose* last position, the weather opened up. They could see for two nautical miles. They were wearing night-vision goggles, 'which

We're Rolling Over

is probably the only way we spotted . . . what looked like a flashing light at the time'. They had already found one life raft and lowered the rescue swimmer into the freezing ocean. It was empty. They left it inflated in case survivors reached it.[36]

In the raft, Dean and Jon could not tell what was happening. After four hours, their suits were iced, and they were freezing and wet. When they saw a light, they thought it was on a boat's mast. There were no flares left, but one of them had the idea to wave a flashlight side to side.

The pilot's testimony: 'And as we brought the swimmer up, the pilot in the right seat who was flying happened to see under his night-vision goggles a waving light, and it was definitely not like the normal blinking light. It was a side-to-side, so we knew it was somebody trying to signal us. So we quickly got the rescue swimmer back up into the helicopter, and . . . at that time, . . . even the flight mechanic was saying that he had to de-ice the rescue swimmer. It was so cold that the rescue swimmer, just from going out the door and coming back up, was covered in ice.' The rescue swimmer was Coast Guard third petty officer Evan Grills. He was 24 years old, he came from Florida, and this was his first rescue.[37]

The pilot had to lower the swimmer safely, but the seas were so wild that the raft kept shifting from one side of the helicopter to the other. The pilot says it was the hardest hoisting he has ever had to do. Evan Grills got ready to drop. 'My nerves were OK,' he told local TV afterward. 'Right up until the door opened.' By the time Grills reached the water, one of the *Scandies Rose* crew had jumped out of the raft into the ocean. I bet it was Dean. He grabbed onto Grills and they were lifted away. Then back down for Jon. By then a Coast Guard aircraft had arrived, and the helicopter pilot chose to take the survivors back rather than stay to search for Gary, David, Brock, Art, and Seth.

In total, the Coast Guard carried out six searches for the missing crew. They spent a further 10½ hours searching 781 square nautical miles and found nothing. Later, the *Scandies Rose* was located and filmed on the seabed by a remotely operated vehicle. Its images showed two immersion suits still in the wheelhouse 'with human remains inside'.

*

In its multiday hearings and subsequent report, the National Transportation Safety Board did not distribute blame for the loss of *Scandies Rose*. The pots should definitely not have been stacked so high and the icing should have been accounted for. The weather forecasting was inadequate and the stability instructions on board were wrong, making the boat more susceptible to capsizing. The report's language is deliberately neutral. 'If vessel captains were aware of the amount of icing.' The cause was determined to be 'inaccurate stability instructions for the vessel'.

A Coast Guard safety booklet based on the *Scandies Rose* sinking was less subtle. Even though Gary Cobban saw the boat had a rapidly worsening list, it stated, 'At no time during his watch did the captain of the *Scandies Rose* declare an emergency for the crew or ring the general alarm to alert the crew, send the crew to investigate the cause of the listing, or prepare to potentially abandon ship.' On the *Scandies Rose*, the VHF radios had a clearly distinguishable red 'distress' button: this needed only one touch and then the boat's information and coordinates would have been broadcast repeatedly and to everyone. But you had to configure the radios first to enable that function, and that had not been done.[38]

In 2020, the survivors and the families of the victims got a $9 million payout from the boat's owners.[39] In 2021, Dean Gribble's fellow survivor, Jon Lawler, died in a motorcycle accident near Anchorage.[40] He had never been the same since the disaster, his sister wrote on a fundraising page, but was constantly 'in a nightmare state'.[41] Dean Gribble has given interviews but not many and usually without the cursing. In Alaska, deaths by fishing have decreased by 70 per cent since 1990 but not by 100 per cent. The missing men of the *Scandies Rose* have not yet been found.

Fishwife at Puerto De La Cruz

5

A Coarse and Vulgar Woman

It's a question of prudence. Nobody has a high opinion of fishwives but who would dare offend them while walking through the fish market.

NICOLAS CHAMFORT[1]

At least half of all workers in the fishing industry are women.[2] They are the processors and the preparers and the cutters and the slicers. They are essential. Without women, you would not have your supermarket prawns or salmon. If women are put in charge of fish ponds, the yield increases. Marie Christine Monfort, President of Women in the Seafood Industry, wrote that everyone has heard of illegal, unreported, and unregulated (IUU) fishing. Women then are IIU: 'the Ignored, Invisible and Unrecognized women in the androcentric seafood industry'. The fishing industry is 'male-dominated but female intensive.'[3] Without women, the fishing industry would collapse.

In 2017, the Mayor of Liverpool was reported to the city council's standards body for making a sexist remark to a female colleague. He was formally reprimanded, and a colleague said of his comment, 'That's just not on. It's not good manners and it's not good politics.'[4] What was the terrible slur he had used? The B-word? The C-word?

He had called Liberal Democrat councillor Mirna Juarez a fishwife.

I look up 'fishwife' in several contemporary dictionaries. Oxford's first definition is 'a coarse-mannered woman who is prone to shouting'. The second definition – 'a woman who sells fish' – is considered archaic. Merriam-Webster decides a fishwife is 'a vulgar abusive woman'. Cambridge has a 'loud, unpleasant woman'. This is its only definition.

Other words for women have migrated from neutral to noxious. 'Virago' comes from *vir* (Latin for 'man'), as does virility, and meant a strong, brave, or warlike woman. Now it is 'a violent, overbearing and ill-tempered' one.[5] 'Hussy' was just short for housewife. The souring of 'fishwife' puzzles me. Why not any street seller? There are plenty in Henry Mayhew's nineteenth-century account of London's itinerant vendors, selling 'walnuts, blacking, apples, onions, braces, combs, turnips, herrings, pens and corn-plaster'.[6] Street sellers were hawkers and not particularly despised. They were a necessary fact of life, however hard that life was. 'It's just another form of starving,' one street seller told Mayhew.

The Mayor of Liverpool could have called his colleague a walnut seller with no sanction. But women who sell fish have always been different. In England they were given permission to sell fish in the fourteenth century, but not in fish markets.[7] They became street 'hucksters', a pejorative word. They were policed as they worked and condemned in legislation. At the end of the sixteenth century, one man earned 20 shillings from the Fishmongers' Company for getting rid of fishwives trespassing in markets. In 1590, a London mayoral proclamation asserted that fishwives were 'not onely of lewde and wicked life and behavior themselves but procurers and drawers of others also servauntes and such like to sundry wicked accions'.[8] Sex and fishwives: the connection was easily made (for if women sell wares, then it is a short step to selling themselves) and stuck. So did the women's working status. When you read about megastar Adele who 'cackled like a fishwife',[9] remember it is code: she is lower class and all her wealth should not let us forget it.

Woman and her fishy smell, her loud voice, her inconvenient presence. Why can't she be quiet?

*

A Coarse and Vulgar Woman

Fishwives weren't always wives. Given their husbands' jobs, often they were widows. Sometimes they were young girls. The 'wife' in fishwife comes from the Old English 'wif', meaning a woman. And they were extraordinary women.

Sir Walter Scott, travelling to north-eastern Scotland's fishing villages, found that 'the government was gynecocracy', a rare term (now only used disparagingly) meaning the political supremacy of women. Women ran things because their men were away fishing. In Scott's *Antiquary*, Mrs Mucklebackit says plainly, 'Them that sell the goods guide the purse – them that guide the purse rule the house.' Elsewhere in the book, a poor country girl expresses pity for the fishwife. You poor drudge, she says. She gets short shrift. How is she a slave, the fishwife points out, when she rules the house because she controls its money? 'Show me ane o' yer bits o' farmer-bodies that wad let their wife drive the stock to the market, and ca' [call] in the debts.'[10]

For these fishwives did everything except go to sea. Instead, they were bait gatherers or fish processors. They walked miles to fetch sand to scatter over their floors, to soak away the smell of the fish they would later dry in their houses. They were fish sellers and fish ambulators, walking hours to sell fish their men had brought home or that they had bought on the quay, heavy creels on their backs, a filleting knife in their skirts. They were merchants and businesswomen. In some Scottish fishing villages, a bride was given a leather purse of money to symbolize this.[11] She was now in charge.

To sell their fish they walked in threes, shifting the carrying. 'It is said,' writes the John Gray Centre in East Lothian, 'that three Fisherrow women once walked the 27 miles from Dunbar to Edinburgh, with 200 pounds of herrings on their backs, in just 5 hours.'[12] It takes me about the same number of hours to run the 26.2 miles of a marathon, and I carry no fish. (Although Sam Shrives Bennett holds the record for 'the fastest marathon dressed as a fisher', and he wore full waterproofs, a sou'wester hat, and wellington boots,[13] the record for anyone running 26.2 miles carrying a laden creel of herring has yet to be attempted.) 'It is perfectly ascertained,' wrote one observer, 'that one, who was delivered [of a child] on Wednesday morning, went to town with her creel on the Saturday forenoon following.'[14]

The observer was the Reverend Dr Alexander Carlyle, who as part of the gigantic parish-by-parish survey that was the Statistical Account of Scotland, in 1792 was given the Parish of Inveresk in East Lothian to report on. There, he found fishwives, and he found that he liked them immensely. His account of their 'peculiar manners' is never scornful. He does not moralize. He records their work, and also their leisure. I will be forever indebted to him for discovering that fishwives played golf. That was not all. 'On Shrove Tuesday there is a standing match at foot-ball, between the married and unmarried women, in which the former are always victors.'[15]

Carlyle admired their strength and also their industriousness, which surpassed even the relentless processing and selling. If there was no harbour, the men had to be got to their boats somehow. In Scottish fishing villages, women carried their men to the boats. First, writes Margaret H. King, they would hitch up their skirts and shove the boat into the sea. 'Then they would carry their menfolk on their backs and deposit them dry shod into the boats.' Otherwise the men would go to sea wet and never dry and probably get pneumonia. In 1917, a medical officer for health in Sutherland wrote:

> The scene at the pier on a cold winter morning is interesting, but somewhat pathetic to a stranger. Women wading through icy water; their petticoats fastened up around their waist, each with her man on her back, and exposed to the gaze of all. To their minds there is nothing indecent in this.[16]

They refused to be photographed doing this work. Only one image is known of, and it is blurred.

Away from the husband-carrying, you couldn't miss a fishwife. The local historians of Sunniside on the north-eastern English coast show magnificent women in Cullercoats dress on their website.[17] Fishwives had a uniform, and it was always local. In Cullercoats, women wore a jacket called a 'bedgown', which wrapped over the front and crossed at the back. Beneath that, an ankle-length skirt, and an apron made 'from any old cloth' to put the fish money in. In Nairn and Newhaven, fishwives wore striped petticoats that were left deliberately visible. Also, 'a black or dark tartan shawl was worn around the shoulders and old stocking legs put over the jumper sleeves to keep them clean. They wore moggans (purses) made either of flannel, canvas, or denim tied round their waists. A moggan had two pockets, one for silver and one for copper. A small purse held pound notes.'[18]

The creel cost good money and sometimes had a lid that could be removed and used as a board to fillet the fish. 'It was usual practice,' wrote the Sunniside Local History Society, 'to deposit the order in the buyer's dish at the door and to fillet the fish. The bones were wrapped up in paper and brought back in the basket at the end of the day. The fishwives were often given cups of tea, or even lunch and many would accept more than they could really manage for fear of showing any favouritism between customers.'

Unseen was the wad of paper or cloth that women put down the backs of their clothing to soak up the wet from the fish. When I learn this I remember meeting manual scavengers – a strange phrase for latrine emptiers – in Gujarat, India, women who had to empty latrines, who carried shit on their heads and also had to come up with protection from drips, but not of water. Those scavengers – who had to empty latrines with their bare hands – were dignified and elegant despite scorn, despite everything.

In the early nineteenth century, King George IV commented on the handsomeness of Newhaven fishwives, and with that they became fashionable. Postcards of them were still popular 50 years later.[19] When the fishwives of Musselburgh went on their annual coach trip in 1939 in their 'gala dress', the local newspaper reported that 'they proved a centre of great interest wherever they went. Cameras were busy . . . [and] snapshots of Fisherrow fishwives are likely to be displayed in other lands.'[20]

*

Sometimes, the women who worked with local fish weren't local but travelling fish processors. They smelled. They were loud. They laughed. They were free. And if you live in the developed world they are gone for good. They were women named after for a humble and essential fish. They were the herring girls.

The herring is a small oily fish. It contains significant amounts of long-chain omega-3 fatty acids, as well as vitamin D. It has, says the MSC, 'been salted in Europe for centuries' and has 'creamy coloured meat, high oil content and a small flake'.[21] Herring – the fish gets that peculiar singular plural that signifies that we are objectifying a living creature – swim in the pelagic zone of the ocean: not near the surface, not in the deep. They can be found in huge schools and can live up to 15 years if we let them. In the 'People Also Ask' section of Google, the top four questions are: 'How long is a herring fish?', 'How big do herrings grow?', 'Is herring a good fish to eat?', 'Do herring have teeth?' (Yes, although only the Atlantic herring.) My favourite query is 'How many herring are there?' because this is a precise version of 'How many fish are there in the sea?', and the contemporary answer is not 'plenty'.

Herring eat copepods and they hunt in packs. There is an animated video of this 'synchro-predation' of 'alert and evasive' copepods, and it is difficult not to think of Pac-Man: the herring swim in three layers, and the copepod attempts to escape one layer then another, but the herring keep coming and they swim with their mouths open and that is the end of the copepods. Smoked, the herring is known as a kipper and is popular in Scotland and among the British upper classes. In 1939, the last year that the herring fishery was allowed to operate off England, it was protected from the air by the Kipper Patrol.[22]

The Swedes ferment herring for six months to become *surströmming*, a product so pungent that a German landlord once evicted a Swedish tenant for spreading *surströmming* brine on the stairs, and the court found in his favour when he opened a can in the courtroom to prove his case.[23] Jamaicans, more modestly, just put chilli on it. Herring is one of the few types of fish caught on a massive scale that is currently doing OK. By 'doing OK', I mean there are enough of them to

A Coarse and Vulgar Woman

be hunted again in great numbers. The herring boom time that began in the mid-nineteenth century is frequently compared to a gold rush. The fish rush lasted longer than gold. Better boats with steam engines and bigger nets allowed shoals to be caught in extraordinary quantities. Sometimes there were a billion fish in a shoal, or there seemed to be (who was counting?). There was such money to be made from the sea that men went off to it, but there was no one then to do the dock work of preserving this fish that was so valuable and so perishable, that rotted so fast. So the women came instead.

They have been called herring girls, no matter their age, or herring lassies if they were Scottish. From about 1850, herring boom time, the herring girls followed the herring as it travelled around the coast of the British Isles, up to Shetland if they were mainlanders, then down from the north to the east coast of Yorkshire, then to East Anglia, and down to the south. The herring travelled in shoals, the women too. In 1909 in Shetland, the herring fishery employed 16,000 people, including 1,670 Shetland women who travelled around the islands to work.[24] Soon they began travelling to England where the fish was and the money. The women did not say it like this. For them, they were 'gyaan sooth tae da' guttin".[25]

They were all ages, and the conditions they endured were often awful. They were transported like cattle on ships sometimes, where they were sick and sick again. The crossing between Aberdeen and Lerwick is still notorious. Up to 16 hours, terrible seas, and they were on deck, probably seated on their 'kist', the trunk that contained their belongings but would also serve as their chairs in the lodging huts. These 'sheddies' were bare, with only wooden bunks provided, along with a roof, walls, and a door: The women and girls had to bring their own crockery, seating, and bedding. Maggie Cowie, interviewed in her old age by the Buckie Heritage Centre in northern Scotland, explained: 'we got a sack and filled it with chaff for yer bed'.[26] Lumpy, and theirs.

Their clothes, said Jeannie Innes, born 1903, were hung on the wall and protected by being covered with tablecloths or old curtains stitched together. 'The trunks had to do for chairs to sit on, there were no chairs!'[27] There was no nothing. Prisoners and fishermen had better amenities than the herring girls.

It was a hard job by any measure, even fishermen's. The days began at six when a cooper – a man who manufactured the herring barrels – knocked on all doors. There was no need for the cooper: the women rose at five to bandage their hands and fingers with 'cloties', scraps of cloth. The Buckie women can still see 'salt holes' on their hands from the combination of salt and water and knives. In the 1920s, the Red Cross of Scotland sent ladies of Edinburgh to minister to the women with aid stations, to dress all the sores.[28] The herring girls worked in teams of three: two gutters and a packer. They stood in all weathers and only ate if there was time and sometimes only bread.

The herring girls did not operate on normal time but on fish time: if there were fishes coming in, they worked. Except on Sunday: the English would work on Sunday but not the 'Scotch'. (This word is no longer used to describe people from Scotland but only for whisky and eggs.) If they got to bed at midnight and a boat came in two hours later, then they went back to work.

The fastest herring lass could gut more than 60 herrings a minute. One second, one fish. They wore whatever was warm and useful: fishermen's jerseys, woollen skirts, oilskins they bought from fishermen friends and cut into apron shapes. They had no shelter in poor weather. It was smelly, cold, awful work sometimes but often it gave them more freedom than any other trade could. The women were a horde and a mass, yet there are photos and postcards of them, dressed in Sunday best, or laughing in the doors of their sheddies, and they are nothing like a mass. You must have had it hard, an interviewer says to Nanie Kaeczmarek of Buckie.[29]

'Och no, it was great.'

In East Anglia, their digs were often summer holiday lodgings run by women. 'Many of these landladies,' wrote Margaret H. King, 'were afraid that the fish-covered clothes of the women would damage their wallpaper, so they hung oilcloth all round the walls up to waist height.'[30] Perhaps the herring women were offended, perhaps not:

when they were housed in plain sheds that they kitted out themselves, they did the same thing, remembering to buy rolls of wallpaper from the 'shoppie' in Aberdeen before they left, then hanging it to hide bare walls, to make a home.

I watch old films that are entitled *Scotch Herring Lassies* and the like. The usual scenes are of the women over the farlins, the troughs that hold the fish to be gutted. They are alike in many ways: in the kerchiefs over their hair, their woollen jumpers, all the sleeves rolled to above the elbow, like healthcare workers. Their oilskin aprons are for practicality but would probably appeal to a sadomasochist. And their hands! So quick. How could they not hurt all the time? All that dipping of their upper bodies and arms into the trough, 60 times a minute. Or the packers, head first in the barrels, rears to the sky like waterbirds.

That one second also had to include triage. In that 60th of a minute, the knife flashed, the guts were removed, and the herring was also sized and assessed, then thrown into the right basket for its size: large full (a herring with roe (eggs) or milt (semen) more than 11.25 inches long), full (same but 10.25 inches), mature full (9.25 inches), medium (a growing herring with no roe or milt), mattie (virgin herring, no milt or roe, more than 9 inches in length), and large spent (herrings that have spawned and are more than 10 inches long). No herring was thrown back, no herring was not fair catch.

In a comment on an old film called *North Sea Herring Fleet*, someone has written:

The herring lasses are hard and tough and fast. No messing. Yet as soon as they have broken every bone in the herring's body to lay it flat in an intricate pattern on the top layer of the barrel, here comes a man with his boots to put the lid on and stamp it shut, banging it with a hammer. As if those tough and fast women in their waterproofs and head-scarfs, smiling at the camera, could not do that with one leg.[31]

Those tough and fast women though: did they have power? I can find no account where they complained of exploitation. The system worked well enough for them. They were taken on by a curer who paid them 'arles', a Scots word that meant an upfront payment. Then they were paid weekly upkeep and got their salary at the end of the season. They had enough money to go to dances and to buy souvenirs. China dogs called wally dogs from Scarborough were a particular hit. Sticks of Yarmouth rock rotted plenty of Buckie teeth. Despite all this, there were strikes, as social scientist Paul Thompson wrote:

> There was a more serious dispute in the autumn of 1936 at Yarmouth, led by Maria Gatt of Rosehearty. There were rough scenes, with mounted police on the streets. Nevertheless, Elsie Farquhar of Buckie remembers that they all rather enjoyed it – and got their wage increase. After World War II there were again strikes, in 1946, 1949, and 1953 – the first winning not only higher pay from the curers, but a more generous coal allowance from the government.[32]

And still the herring girls lost their jobs when the herring disappeared because of overfishing. The itinerant fishwives in coastal villages stopped selling fish when it began to be sold instead to shops and then supermarkets. This does not mean that women do not work in fishing, or that they do not hold up the world in fishing as in everything else. You just don't see them. They have aquaculture ponds or they fillet fish in dusty, smelly markets. They work in canning factories and pet food facilities. They live on enormous factory freezer trawlers for many months, working below deck heading and gutting. They process and transform, and they are as essential as they always were. In industrialized countries in modern times, they have retreated to factories, indoors, out of sight.

Except for Big Lil.

*

A Coarse and Vulgar Woman

St Romanus, 20 on board, all hands lost.
Kingston Peridot, 20 on board, all hands lost.
Ross Cleveland, 20 on board, 19 hands lost.

The *St Romanus* went first. Then the *Kingston Peridot*, then the *Ross Cleveland*, within 30 days. They were trawlers from Hull, a city set on the bank of the great River Humber, whose generous mouth leads to the sea, a superhighway to distant water. In 1968, Hull had one of the most substantial distant-water fisheries in the world. About 150 trawlers fetched a quarter of the fish that Britain ate.[33] The pay and the fish were so good that Hull trawlermen were known as three-day millionaires for they often had only three days at home, and the unmarried crew, flush with wages, spent those days as unwisely as possible. The fishing was good because of where it was: a thousand miles away beyond safety and the coast. Hull's trawlers still went out to the best fishing grounds off Iceland, in any season. British ships had fished off Iceland since the fourteenth century and off Newfoundland in Canada since the seventeenth century. They were used to weather and trouble.

The start of 1968 brought more than the usual of both. The severe months of January and February, the months that sneak in with their storms and fury under the relief of a new year, were devastating for Hull. First the *St Romanus* went silent. Hi-vis and health and safety regulations did not exist in 1968. It was accepted that more deep-sea fishermen died than coal miners[34] and that more than 5,000 fishermen from Hull had died on the job in 100 years.[35] Hardly any trawlers carried a radio or someone to operate it. Lifejackets? Nay, lad. They made the fishing work impossible. Fishermen, it was said, 'are men who work with death at their left hands'.[36] The trawlers went out there, and they stayed out there, unhitched from the land until they came back. The *St Romanus* was last heard from on 10 January, but the alarm was raised only on 26 January. The owners' justification was that skippers often went silent when they had good fishing, so that competitors would not head for the same fishing grounds.

On the day the alarm was raised for the *St Romanus*, the *Kingston Peridot* radioed that there was ice build-up on the boat. That was the last message from the boat, and a life raft washed ashore the next day. By 30 January, the *Kingston Peridot* was reported lost.

Two boats lost, 40 men dead. The press called Hull 'The Sad City'.[37] And finally the women of Hull thought that it should not be fishing business as usual.

Lillian Bilocca was what is called a fishwife. She worked in a fish factory skinning cod; her son was a trawler deckhand, her husband a merchant seaman now ashore. She was sea people. There are pictures of her in the fish factory that show why she was called 'indomitable' or 'fierce', or why her nickname was 'Big Lil'. She weighed 17 stone and wore her rubber filleting apron like armour. I prefer the adjective 'magnificent'. A big woman, big hair, big earrings. She looks terrifying and fabulous, like the best dinner lady you can remember from primary school. Lil's daughter remembers her thumping the table when Lil heard on the radio that the *Kingston Peridot* had gone too. 'Enough's enough,' she said. And she became a woman 'with anger on her lips', as a male journalist wrote, and a headscarf on her big black backcombed hair.[38] Big Lil and three other fish women became the Headscarf Revolutionaries because they wanted change and working-class women wore headscarves in those days. Lillian, but also Mary Denness, Yvonne Blenkinsop, and Christine Jensen. They wanted better safety, radio operators, better training. They wanted their men to stop dying because the boats they fished from weren't good enough. The women knew that weather and ice could always tip a boat, that no vessel is unsinkable, but why give the sea a hand by sending men out without radios that can call for help, or alarms in their cabins? Trawlers had radio rooms, but they rarely carried enough crew to man them. Instead, there was usually only a radio in the wheelhouse, which may or may not get used.

The women drew up a Fishermen's Charter. This listed their demands: improved weather reports and more safety equipment. A 'mother ship' that sailed with the fleet and had medical facilities. European fleets had such things; why couldn't their men? They wanted an end to fishing in north Icelandic winter storms. They wanted vessels to report in every 12 hours without fail. They wanted an end to 'Christmas crackers', poorly trained last-minute crew employed around Christmas.[39] They wanted trawler owners to value the men who fished as much as they valued the fish. Every year, the Hull trawler owners' association awarded the Silver Cod Trophy to the trawler with the largest catch of cod. To catch the best catch, trawlers had to go to the most dangerous grounds in the worst weather. They had to fish no matter what. The trophy encouraged risk.

First the women had a meeting in Victoria Hall. It was packed with mostly women and many headscarves. Then they founded the Hessle Road Women's Committee, named after the notorious street near the docks where most fishing families lived, and started a petition. The petition got 10,000 signatures in three days in an area with a population of 14,000.[40]

There was door knocking and rousing, and there was drama. In the first days of February, Big Lil went to St Andrew's Dock. This was a courageous thing to do in a superstitious fishing town. Women didn't go to the dock, they didn't see their men off, and they certainly didn't stand on the quay yelling at trawlers, 'Got a full crew, lads? Got a radio operator?' If a crew said yes, they did, she shouted, 'All the best, flowers.'[41] The first trawler crew to say they had no radio operator was from the *St Keverne*. At that point Lillian erupted. She had said she would launch herself onto any trawler that had no dedicated radio operator. This, as Brian Lavery, author of the fine book *The Headscarf Revolutionaries*, points out, would probably have killed her, but never mind that. She was stopped by a few police officers, but she kept struggling. She yelled at the trawler heading out to the river and then the sea. 'Don't go without a radio operator, lads! Don't go! Don't go!' And though the trawler had set off regardless, 'the *St Keverne* sailed through the lock gates into the River Humber. Then it circled – and stopped. A dramatic message was sent ashore . . . "The crew will not sail without a radio operator.'"[42]

The petition was successful, and the women were invited to Westminster to meet officials from the Board of Trade. As Lavery writes, BBC reporter James Goodrick interviewed Lillian on the station platform as she waited for the 07:55 to King's Cross. He asked her what she was planning. 'If I can get satisfaction from the ministers I am prepared to come home and carry on but if I don't get satisfaction I'll be at that Wilson's house [Harold Wilson, then prime minister], that private house, 'til I do get satisfaction in some shape or form.' 'Oh,' said Mary Dennis, 'she's off on one now!'[43]

The meeting was almost comical. These working-class women from Hull's fishing streets, and a battalion of men in dark suits. Whisky on the table, smoke in the air. The women had no time for deference. Bilocca said, of the ministers, 'We saw this bloke Mallalieu [Minister of State at the Board of Trade] and t'other thing Peart [Minister of Agriculture and Fisheries]. I told them: "Speak my language and don't use a lot of your long words," and they did speak our sort of language.'[44]

Yvonne Blenkinsop called Mallalieu 'petal'. Two classes, the highest and the lowest, meeting like waves. 'When we was coming out,' Yvonne Blenkinsop told the BBC, 'I said, "Petal, are we going to have these things then?" And he said, "You are, my dear." Real nice, with a big smile. He agreed with everything all of us was saying because they all needed doing. Everything. Every one. Now that's good.'

The press loved these fishwives. The *People*'s headline was 'SOBBING WIFE BRINGS TRAWLER VICTORY'.[45] *The Times* described Yvonne Blenkinsop as 'the epitome of a fishing port wife' who stood 'arms akimbo,' and 'whom strong men would confront at their peril'.[46] Not all fishermen loved them: many thought their jobs were being risked, and Blenkinsop was punched in the face by one trawlerman when she went out for tea with her husband. 'It was a right wallop.' They all got death threats in the mail and were briefly given police protection. These uppity women: how dare they? Letters arrived for Lillian that at first she burned and then she kept to show the police. One read, 'Madam,

A Coarse and Vulgar Woman

Why don't the people of Hull kidnap you, tie some bricks round your neck and drop you in the Humber, you big, fat, greasy Maltese whore. You must be the commonest cow in Hull. Ask your xxxx Maltese husband to take you to Malta and lose you.'[47] She even got telegrams. Fancy going to the bother of that: to take your hate to the post office and write a telegram and pay for it. I hope Lillian got a laugh at least at the one that read YOU SHOULD GO TO PRISON ON A SLIMMING DIET YOU ARE VERY COMMON AND HORRIBLE.[48]

These women with their ministerial meetings, with their uncommon reach: they were still common fishwives. Their place was the shore not the ship, the kitchen not politics, quietness not noise. They had no business on trawlers nor speaking for trawlermen. Michael Burton, president of the Hull Fishing Vessel Owners' Association, told an interviewer, 'I have much more sympathy with the relatives who've been lost at sea, frankly, than with a lot of women who are trying to or not trying, but are getting carried away on a wave of mass hysteria.' And then: 'Frankly, the ordinary fisherman is a bit sick of all these women, interfering in their own business. And the sooner we get down to dealing with the men who matter rather than the women, the better.'[49]

On 5 February, the day the headscarf women met trawler owners, and the day before they met ministers, the *Ross Cleveland* sank in Ísafjörður fjord off Iceland in a terrible storm. The ship's mate, a 26-year-old Hull man named Harry Eddom, was working on deck chipping ice from the radar. He remembers that it took six or seven seconds to go from upright to capsized. There was enough time for skipper Phil Gay to radio, 'We are laying over. Help me. I'm going over,' and, 'We're going. Give my love and the crew's love to our wives and families.'[50] Eddom was thrown into the sea and awoke in a rubber life raft with a man and a boy. The boy was dressed in a T-shirt and underpants. They bailed water with their boots but only Eddom's outdoor deck-work clothes gave him any protection. He watched the boy die and then the bosun, and 13 hours later was washed up to the Icelandic coast. He and his two dead crewmates. There he began walking and after nine hours found a hut.[51] He sheltered behind it, as his legs were too frozen to kick the door down. He stayed awake and standing all night because, he told the *Daily Express*, 'if I had sat down I would have died'. In the morning after what must have been a horrific night, a young shepherd passed. Eddom managed to call to him, and 'he ran over, put his arm round me, and helped me to his father's farm'.

The *Ross Cleveland*'s 19 men were presumed dead. Lillian Bilocca's son Ernie was on the *Kingston Andalusite*, part of the same fleet, facing the same weather. Ernie watched the *Ross Cleveland* disappear from the radar screen in the *Andalusite*'s wheelhouse. While the women were at their London meeting, they were told that Harry Eddom had survived the sinking. Big Lil cried at a press conference when she heard the news. She said, 'I felt fifty instead of forty when I came down here today but I feel happier now.'[52]

After three weeks of campaigning, Bilocca was sacked from her factory job and she was certain that the trawler owners were behind it. By this time, she was enough of a celebrity that the news of her sacking appeared on the front page of *The Times* underneath a story about Indira Gandhi and a picture of Viscount Linley arriving for Prince Andrew's eighth birthday party.[53] The most famous woman in fishing was interviewed by Marjorie Proops, the most famous woman in journalism. Proops called her 'the suffragette of 1968' and wrote that 'I had expected aggression but I found sorrow.' Proops asked why grown men doing dangerous jobs need to have women sticking up for them. Because they sit around and 'do nowt', said Lil. 'They haven't got the time, or the will.' Proops liked Lil. They were of a kind. She wrote of a woman who had 'made a 17-stone dent in complacency'. She wrote that 'someone ought to erect a 17-stone statue of Lil on the fish docks at Hull. It would be a real Statue of Liberty.'[54]

Lil finally found menial work after two years of looking. This was never headline news. How useful was her community in helping her? As useful as the authorities had been in improving safety conditions for fishermen. As useful as a sugar fishing net. At the time, wrote one ex-seafarer, the industry was 'almost proud of its abysmal safety

A Coarse and Vulgar Woman

record'.[55] In a parliamentary debate, Charles Fletcher-Cooke, the Conservative MP for Darwen, had said, 'Trawlers seem to be going down like ninepins, and one is prompted to ask why.'[56] The mildness of this question is infuriating, and without Lil, it would have sunk as quietly as any inquiry into the riskiest job in the world usually did. There had been trawler tragedies before, when in 1955 the *Rodrigo* and the *Lorella* capsized on the same night because of icing.[57] Nothing was changed. Seven years later, the Board of Trade decided that the installation of alarms in crew cabins, so that they were alerted in good time if their ship was listing or sinking, should be optional.

For anyone who understands how slowly government moves, the speed of the changes to trawler safety will be shocking. On 8 February, the same day that Proops' 'the Real Big Lil' interview was published, the Board of Trade announced 88 new measures. All trawlers would have a dedicated radio officer who was not the skipper. A mother ship would be in operation within a week. All the trawlers in the northern Icelandic waters were recalled immediately and no fishing off Iceland would be allowed until there was better weather. The mother ship would be provided by the Board of Trade. The first was the Hull trawler *Orsino*, converted to both mother ship and weather ship, with a crew of 20 men including a doctor, and meteorological equipment. It set sail in November.[58]

Despite this, people blamed Big Lil for her efforts. Astonishingly, they also blamed Harry Eddom for surviving when 58 didn't. It is astounding what criticisms people can dream up: he should have given his suit to the young boy in the raft. He should have been dead because everyone else was. Eleven weeks after he got home from Iceland, he sailed as first mate on the *Ross Antares*. He was escorted to his trawler by the police.[59]

The Headscarf Revolutionaries did not burn bright for long. Hull's distant-water fishery collapsed in the 1980s. Iceland closed its waters, the bitter Cod Wars between British and Icelandic trawlers were won by the Icelanders, and Britons now eat Icelandic fish with their chips.

In 1988, Lillian Bilocca died of cancer at the age of 59.[60] Hardly anyone outside her family attended her funeral.

Hull is a different city now. It still claims a maritime identity. It has a well-attended aquarium and a maritime museum. Hessle Road is drab, and what is left of its fishing community has moved away to distant housing estates. St Andrew's Dock has no trawlers and is a wasteland waiting redevelopment. I have come to Coltman Street off Hessle Road. At this end, it was a skippers' street: the houses are bigger and the road is wider. Further down, narrower and shabbier, the street housed crew. George Rose, the drowned 35-year-old third mate on the *Kingston Peridot* and my backward namesake, lived at number 142.

Big Lil is getting a blue plaque. There is already a mural honouring Lillian and her fellow revolutionaries, but this blue plaque is a big deal and there is a throng. It has been fixed to the house where Lillian lived, a square-fronted place that looks both grand and shabby, now home to a young family. There are TV cameras and furry mikes, and dozens of women in headscarves in tribute. I ask a young woman next to me what her connection to Big Lil was. 'She was my grandma.' I had been reading about her dad, Ernie, the night before. The granddaughter said she used to come and play at the house, and being called Bilocca was tricky when she was trying to get into pubs underage, because everyone knew Ernie's lass, and everyone knew Big Lil.

There are speeches from a gold-wearing Lord Mayor and other dignitaries, and someone wins a portrait of Lillian. I get talking to a man who is a local councillor, and a Chilean. How does a Chilean get to be a Hull councillor? Herraldo says simply, 'Pinochet.' I ask him what the purpose of a blue plaque now is. He says, there aren't many working-class heroes, but Lillian was one of them. He says that he can meet children in school and tell them about Lillian and they can think that they can do that too. That's why this matters. I tell him that is a very good answer because it is.

Perhaps a blue plaque is a ripple too small for someone of the stature of Big Lil and the fishwives and fishers' wives who did more in a month to save lives at sea than the unions and government authorities had achieved in decades. But it's a start. A campaign to get a statue of Big Lil is ongoing. In 2021, Hull residents were asked to vote for the Greatest Hullensian of all time. Lillian was one; another was pop musician Paul Heaton; and another was the eminent anti-slavery campaigner William Wilberforce. Wilberforce won.[61]

'I've always been concerned,' Lillian said soon after she launched the campaign. 'But I've never had the guts to do owt about it, but now it's time somebody did and I've made a start. It's up to other people to follow me and make these owners sit up and take bloody notice.'[62] Swearing, loud, opinionated, magnificent, working-class: Big Lil gave fishwives the good name they have always deserved.

Captain Tom's wheelhouse

6

Trawling

*We caught everything we saw. The companies
wanted to catch everything but the wiggle.*

HALL WATERS, FISH SPOTTER PILOT[1]

'Pillaging the seabed, that's what we're doing,' says Captain Tom McClure on his trawler *Guardian*, a couple of hours off Newlyn in Cornwall. Tom has little time for this accusation from environmentalists: he thinks it is unrealistic and unfair. The eco people in turn have little time for bottom trawling, and any kind of fishing gear that rakes the seabed, because they think it indiscriminate, damaging, and wasteful.

Tom knows all this. Still, he thinks his fishing is sustainable and nothing like pillaging. The seabed recovers, he says, and he should know: he is out here all the time, except for the last six months, when his last boat failed and it took him all that time to complete the paperwork to get his new one going. The new boat cost £420,000 ($525,000) from a fisherman in Looe, and he was pleased to get it at that price because it's only eight years old. The *Guardian* is not perfect: the winches are in the wrong place, there isn't enough room to remove fishes from the net, and he would like an on-deck radar monitor so that he doesn't have to keep stepping back into the wheelhouse with his fishy boots to check what is what. At least he is back at sea.

I first encountered Tom on TV, in the BBC series *A Fishing Life*, one of the few documentary series about fishing that is not big men on big boats catching big fish or big crabs. 'Ah,' he said on the phone when I called him and told him I'd seen his episode, and repeated what he had said on camera, unexpectedly. 'The lucky testicle.' He had said that to wind up the cameraman and hadn't thought they would keep it in, but they did, and now it is all that people remember of his starring role.

When Tom had invited me fishing the day before, he told me departure would be at 4 a.m. The French call the hours between dusk and sunset 'between the dog and the wolf', but they probably don't have an animal for what the world feels like at 3 a.m., sitting at a borrowed kitchen table, trying desperately to eat something to line my stomach though not too much, because this expensive boat was designed without a toilet. A message from Tom: 'Just to remind you the toilet facilities are bucket and chuck it.'

I had packed for what I knew from fishing with Rex: many hours on a small boat with no shelter. Oilskins, boots, hat, gloves, extra fleeces. When I arrive on a dark dock, both Tom and his crew member, a former taxi driver also called Tom, are in Crocs and trainers and casual trousers and shorts. This is a luxury boat compared to Rex's crabber: there is a downstairs with four bunks, a microwave and oven, and a kettle. The interior is warm and cosy. Young Tom is brand new. He has done some stints at sea but not on a bottom trawler. When he sees me, he says, 'Are you fisheries?' Fishermen are used to and sick of having visitors on board. Camera crews, fisheries observers, marine biologists, people with clipboards. I'm not certain that 'No, I'm writing a book' is less intrusive, but young Tom asks no further questions.

First, we will steam: a few hours to somewhere offshore before any work has to be done. The work will mean noise: the winches on the back of the boat that lower the nets into the sea, then the 'otters', two heavy metal boards that are lowered last and which trail in front of the net like hydroplanes, holding the net open. I ask Tom how long the nets are and get this in response:

'The length of the front is 15 fathoms. It won't be 15 fathoms, 90 feet straight down. It's a U-shape, so probably only about 40 feet wing-tip to wing-tip.'

I'm not sure where to start so I try the obvious. How big is a fathom? I had no idea fathoms could be a linear measurement. Dumbly I thought of a fathom as only depth. Which is a linear measurement. Tom sees my cognitive trouble and helps me out. 'It's like a big sock.' It will be dragged along the seabed for 4½ hours. Why so precise? 'That's just what works. Too short and you don't get a good catch. Longer than that and the fish get damaged.'

The noise is mechanical, but beyond that there is mostly quiet. You don't shout on a fishing vessel because shouting means something is seriously wrong. A good skipper does not need to shout; a bad one chooses to. Before the first haul, there is little business to be done: one Tom goes down to sleep, the other stays in the wheelhouse. This has two seats and a large counter that holds the radar, sea plotter, and the sounder or fish finder. It also has an automatic identification system (AIS) that transmits his AIS number, which can be switched off. Fishing vessels under 24 metres are not required to have an International Maritime Organization (IMO) number, a permanent ID, and the *Guardian* is only 10 metres long.[2] It is a small boat but the wheel is bigger than the one I saw on the Maersk *Kendal*, a container ship 20 times the length of the *Guardian*. That was a pathetic joystick. Tom's previous boat had a wooden wheel.

The sophistication of the machinery on fishing vessels increases with the scale of the fishery and its possible profits. Tuna finders use helicopter spotter planes. Tuna boats also have fish aggregating devices (FADs), anchored or drifting structures – often wooden rafts – used to attract smaller fishes (who like to congregate around any object in the ocean), which then attract tuna. They also attract five times more by-catch than any other fishing method. There are at least 100,000 FADs in the oceans, but 60 per cent are lost, abandoned, or stolen according to the Marine Conservation Institute.[3] Used or disused, they still attract and trap creatures in their nets and ropes.[4]

Tom doesn't use FADs. He doesn't use a plane. He uses a big sock, heavy otter doors, and 30 years of experience.

Tom has white curly hair and a Cornish accent that softens his comments. I laugh along with him when he makes his cracks about pillaging the seabed because I like him and he has allowed me to fish with him on his new boat. He is kind each of the four times I throw up over the side of the *Guardian*, once in the middle of a conversation with him in the wheelhouse, when my stomach started swimming and all I could do was make the sign of stop with my finger, point to my stomach, make a swirling motion, then dash to my usual spot, on the wrong side so I had to vomit over the fenders. This is a basic error: splashback is more likely when you puke over giant rubber tyres attached to the side of the boat. I should have known better. I am an expert at vomiting.

I have had motion sickness all my life. I watch people reading blithely in cars and I am envious. My limit is three words, and a map is out of the question. I am carsick and bus sick but not train sick or plane sick. I was sick when I had surgery (and was told by an anaesthetist friend that I had several risk factors: my surgery was abdominal, I'm female, I don't smoke, and I was given opioids). I am always boat sick. I once turned pale on a cruise ship during a comedy show and the comedian publicly mocked me for looking miserable. She apologized the next morning at breakfast, and said, 'I saw you were green; I shouldn't have done it.'

I did not turn green. A faint wave in the stomach, an uncomfortable warmth. These symptoms have a name: they are 'prodromal', the signs of the sickness to come. An interval and a signal that everything is wrong. (Probably we say green because 'white' skin changes from pinkish to a paler colour. Only Caucasians are said to turn green.)

In January 1842, Charles Dickens crossed the Atlantic Ocean on a steamer. A century and a half later, I followed. His ship was the RMS *Britannia*; mine was the Canmar *Pride*, a midsize container vessel. My ship was bigger, but the weather was the same size and the horizon equally crazy. On the third day of his voyage east, Dickens woke to find the stateroom mirror on the ceiling and the ship 'staggering, heaving, wrestling, leaping, diving, jumping, pitching, throbbing, rolling, and rocking; and going through all these movements, sometimes by turns, and sometimes altogether; until one feels disposed to roar for mercy.' He asked a steward what the matter was. 'Rather a heavy sea on, sir, and a headwind.' This was the beginning of his seasickness: a new affliction, for him, that sank him into a lethargy close to death. (The old seaman's joke: the worst thing about seasickness is knowing you are not going to die.) The lethargy was also not the ordinary kind, but a profound kind of nonliving where, 'If Neptune himself had walked in, with a toasted shark on his trident, I should have looked upon the event as one of the very commonest of everyday occurrences.'[5]

Seasickness is both ancient and modern and as unresolved as it was when ancient Chinese scholars wrote of 'cart-sickness' and 'litter-sickness'. Remedies have been varied and vivid, as Doreen Huppert and colleagues wrote in a history of 'a Plague at Sea and on Land and also with Military Impact'.

> The western classics recommended therapeutic measures like fasting or specific diets, pleasant fragrancies, medicinal plants like white hellebore (containing various alkaloids), or a mixture of wine and wormwood. The East knew more unusual measures, such as drinking the urine of young boys, swallowing white sand-syrup, collecting water drops from a bamboo stick, or hiding earth from the kitchen hearth under the hair.[6]

There is no remedy for seasickness except to stop going to sea. In the Royal Navy, sailors who suffer the worst are assigned to the bigger ships, which have more stability. Or they stop being sailors. In fishing, seasickness is common enough not to be scorned, but also uncommon because if you don't get fixed, you don't go fishing. I don't want my stomach to stop me from going to sea. I take seasickness pills, which don't really work. I take a bag of crystallized ginger and acupressure bands. I pack doughnuts to line my stomach. I find online that there are now goggles available, but they look ridiculous and I know I'd be laughed off a boat. I go to sea knowing I will heave, as sure as the tide goes in and out.

Cats are common subjects for motion sickness studies: they get sick and they are available. However, shrews are also commonly used, and are often attached to a 'table-top shaker' in the cause of furthering human knowledge.[7] Rabbits don't vomit, nor do horses or rodents. ('Can Rabbits Throw Up?' is a seam of the internet I did not expect to enjoy.) You may feel sick as a dog, but which one? Some dogs are susceptible; others are immune. Fishes get seasick, at least according to one German scientist.[8] If sheep couldn't be sick, the livestock transportation business would be a lot happier. The sheep-shipping industry that sends thousands of Australian animals to the Middle East by sea is beset by dead sheep, who usually die because they stop eating. There are theories about their mortality rates. They don't like the food pellets; they don't like the high ammonia levels in their environment; they have metabolic diseases. No way. It's because they feel like I do at sea. But it has taken science to show that.

In 2015, scientists at the University of Queensland chose to test sheep's sea legs by simulating a vessel's roll and heave. They restrained six Merinos in a crate, in pairs, and placed them on 'a moveable programmable platform' that created the roll. A forklift created the heave. For an hour a day for 12 days, the sheep heaved and rolled so that humans could understand whether their feeding was affected. Some sheep were given anti-emetics; some were not. The scientists learned the following: that sheep in the heave group ate faster and also put their heads more often against the mesh of the crate, 'perhaps for balance'. The ones given anti-sickness medications rested their heads less on the mesh. 'It is concluded,' wrote the authors, 'that simulated ship motion had adverse effects on feeding behaviour and balance, which appeared to be attenuated by antiemetics.'[9] Twelve hours of nausea. I felt for the sheep.

The science of researching sickness in humans is more advanced, but the principle is the same: on land, a human is strapped into a chair and it is spun in a way that will confuse the brain, though not like being on a ship will. When I am sitting below deck on a ship, my brain thinks I am still. The ship movements tell my otoliths – tiny bones in my ears involved in regulating gravity and movement – something else. This is well understood. Less understood is why this would cause me to discharge the contents of my stomach over the side of Captain Tom's boat, four times in 13 hours, until I am heaving up nothing but still heaving. Mary Roach, in her marvellous *Packing for Mars*, an account of what is required to send humans to space, quotes a NASA scientist who calls motion sickness 'one of God's jokes'.[10] In the brain stem, there is an emetic centre (also known as the vomiting centre) and also a small area that detects chemical toxins and disturbance. The vestibular system, which governs our balance and keeps us stable on a planet moving through space, is right next to it. So disturbance to one sets off the other.

As I write this, on dry land at a desk, my throat contracts as it does before I feel sick. This is because merely writing about being sick is 'provocative' to my emetic centre even when my vestibular system is content. The sea, in motion sickness research terminology, is a 'provocative environment' even for fish. Whether fish get sick has been a scientific mystery for a while now, but in 2009, a German scientist thought he had cracked it. Reinhold Hilbig sent 49 fish in a mini-aquarium up in a plane that then dived steeply. This was supposed to mimic the loss of gravity in a spaceship. Eight fish started turning in circles, said Hilbig, proving that they were seasick. 'They looked as if they were about to vomit,' he said, unconvincingly. For now, Hilbig's eight seasick fish are on their own in the literature: A search of 'motion sickness' and 'fish' only yields countless pieces of advice for people setting out on fishing charters.[11]

For the most money thrown at motion sickness research, you used to have to head for space medicine. These days, the sickness money is in driverless cars and virtual reality. Fishermen and their queasy passengers just get used to it or stay ashore.

*

I soak up Tom's sea wisdom, because it includes knowing that you cannot know anything. 'The bad fishing,' he says, 'is just one of those things. May, June can be contrary, I don't know what it is, the change of temperature, the daylight. Fish are quite susceptible to weather, and we've had a lot of easterly and north-easterly winds, and they don't like that. There has also been a lot of mild weather and the fish like to be stirred up once in a while.' I ask Tom where he learned to read the winds and the sea. 'It's just what you pick up, over the years, chatting to different people, observations. There's no classroom where you go and learn this knowledge, it's what you acquire.'

The skipper's seat is the right-hand one. From there, he can see everything and reach the radio above his head on the right. There is debris on the sloping counter, enough that it looks like my kitchen table. An empty mug with tea dregs, the *Guardian*'s logbook, bits of paper, magazines, odds and sods, and bits and bobs. The wheelhouse is warm, and it is not the open deck with its noise and smell and weather, so this is where I stay. Anyway, when the trawl is hauled in, I won't be allowed on deck. It is too hazardous for a visitor. It is equally difficult for a fisherman – a rolling boat, heavy machinery, winches under extreme tension – but they are supposed to be used to it.

The trawl is still trawling, so there is time to talk. Tom was not always a trawlerman. He is not a usual fisherman to begin with. His family is from Penzance, but they did not fish. His father was a dentist; his mother was fiercely opposed to her son fishing. A childhood 'mucking about in the harbour', then a boat design course in nearby Falmouth he didn't enjoy, and 20-year-old Tom McClure was back in Penzance and back to sea. He started on a gillnetter, a boat that trails a wall of netting. A gill net can be five miles long and is designed to trap a fish by its gills in the mesh.

After that, Captain Tom was two years on a deep-sea trawler 36 metres long. Then he bought the *Harvest Reaper*, a 10-metre fishing boat that served him well. The newer *Guardian*, he calculated, would serve him until retirement. He doesn't expect to fish forever. By the time the *Guardian* was ready and legal, he had exhausted his savings and now he was starting to fish just as the price of everything went nuts. Fuel, ice, everything. One of his common sayings is, 'Bloody Putin.'

The trawl is done and the haul now begins. I stay in the wheelhouse while the two Toms get to work on deck. One operates the winch; the other checks and empties the net. The work looks hard and wet even on a good weather day. I am transfixed by the starfish in the net, their limbs sticking out of the mesh, still moving, forlorn. Starfish are considered trash fish: of use to no one except themselves and the ecosystem and marine food web. I come to think of them as the most symbolic wastes of trawling. The prettiest, the saddest. My dinner-table fact about starfish is how they capture their prey such as scallops. 'They're all tube feet,' a marine biologist tells me, 'so it's just like a slow powerful pull, then the starfish feed by exuding their stomach into the cavity of the scallop, pumping out digestive enzymes so it digests things externally. Don't get eaten by a starfish, it's not a very nice way to go.'

There is little dignity for the unwanted net-caught creatures: They are dumped in a small area near the stern and the Toms pick through them. Keep this, throw that back. Keep that, throw this back. This is money, this is not.

Some things that fishing gear catch as well as the target fish: turtles, porpoises, dolphins, sharks (silky, catsharks, makos), whales. Biogenic life such as polychaetes, coral, shellfish. Gannets, gulls, kittiwakes, Balearic shearwaters. Pregnant fish that should not be caught and definitely not landed. Young fish that should be also left in the sea to grow into adult fish. Endangered rays (devil rays, stingrays, blond rays, Atlantic torpedo rays). The International Council for the Exploration of the Sea produces a roadmap of accidentally caught species divided by region. The list only focuses on protected, endangered, and threatened species and is still a depressing 36 pages long.[12]

Every fishery management organization and fisheries governance department has reams of pages devoted to reducing discards and by-catch. By the amount of paper produced to solve the problem, you would expect it by now to be as small as a starfish.

In an introduction to a report on wildlife by-catch, John Tanzer of WWF wrote: 'Bycatch is the main driver of decline and threat of extinction in a number of endangered or critically endangered marine species, and staggering statistics estimate that every year fisheries by-catch kills: 720,000 seabirds, 300,000 whales and dolphins, 345,000 seals and sealions, over 250,000 turtles, and more than 1.1 million tonnes of sharks and rays.'[13]

In 2021, Stella Nemecky published a paper on discards.[14] She is a marine biologist at WWF Deutschland who was needled into writing her paper by the slow progress of the European Commission's reform of its Fisheries Control Regulation. The European Union announced this reform in 2018. By 2021, it had still not been published, so Nemecky got to work. One of the proposed measures was to require fishing vessels to carry remote electronic monitoring, CCTV on boats that would reduce the number of undeclared discards. Currently all data on discards is self-reported. Self-reported must mean under-reported: why would you make yourself look bad when no one is watching?

'Let's say you have a quota for cod,' says Nemecky, 'and that's going to be the smallest quota you have because the cod stocks are in a mess, and you catch your quota. If you carry on fishing to fill your other quotas for other species, you're bound to catch more cod.' What do you do with it? You either have to stop fishing and miss out on your other quotas, or it goes overboard. And that is a death sentence for most fish. Flat fish have a better survival rate, but depending on the fishing gear used, most fish will be too traumatized to survive. 'If they're caught in a net,' says Nemecky, 'they will continue to swim. It's like they're running running running, there's too much air, too much pressure. It's often a combination of stress and getting crushed and it also depends on what depth they were pulled up from.' From certain depths, being hauled up the water column would put huge pressure on the swim bladder, an internal organ filled with gas that works like a buoyancy control device for a diver, only much better. A fish suffering from barotrauma – injury caused by a change in pressure – could have its stomach protruding through its mouth, or bulging eyes, or a bloated belly. There are simple solutions (apart from not fishing). The headline on one blog post reads 'SWIM BLADDER: WOULD BEING STABBED BE BENEFICIAL TO A FISH?'[15]

Yes, it would. The technique is called 'venting' and is done with a venting tool that can be bought for under $10. The job is stabbing, and it requires a sharp steel tube that is hollow to ensure a wide enough aperture for all that gas to escape. Venting tools are marketed to anglers and sports fishermen. No commercial fisherman will have the time or inclination to vent the stomach of every barotraumatized fish.

The fishing industry refers to 'choke species'. These fishes are known by fishermen to be so abundant that they will inevitably be caught in any mixed fishery but are subject to controls and quotas, which many fishermen think are to be out of date and unrelated to reality. A translation: we are not allowed to catch cod, but there is so much we can't not catch it, and that's better than chucking it dead overboard. To land such quota species when you have filled your quota is to land 'black fish', and it is illegal. In 2009, six boat owners in Newlyn were prosecuted for landing black fish (hake that was passed off as turbot, for example) and given criminal records, even though two were elderly women over 80. One was 83-year-old Joan Turtle, a part-owner of a fishing vessel skippered by her son James. 'I cannot believe,' she said, 'I have ended up in court as a criminal when all I have done is tried to support my husband and family. Can you imagine throwing one-third of your wages in the dustbin? It was need not greed. The gut feeling of having to throw dead healthy fish over the side when you could not avoid catching it – I can't imagine how it was.'[16]

Some fishes have a better chance than others when they are accidentally caught. Plaice, says Nemecky, will be OK but more so in winter with careful fishermen. 'If they're fished in winter with a short haul and they're treated well on board, so they're handled quickly and fairly cold, survival rate is high. If you have a different net during summertime, and the haul takes a long time and it's really warm, lack of oxygen and so on, the survival rate decreases massively.'

Skipper Tom's problems are spur dogs, another name for smooth hounds, a type of shark. They swim in shoals and sometimes they swim into Tom's net. 'I can't land them,' he says. 'About 30 years ago, they were overfished, no doubt about it. They have recovered in the last couple of years. But if they go in your trawl they bung up your trawl and they're a nightmare to deal with.' A shoal can get caught in the trawl and then Tom must throw them back. Gillnetters, whose nets hang static in the water, can land them but boats with mobile gear such as a trawl can't. Like the starfish and the rest of the unwanted, they go back to the sea alive but unlikely to survive.

Tom can only land fishes that he has quota for. He gets his quota from the Cornwall Fish Producers' Association, of which he is a paid member. Small-scale British fishermen must belong to a producers' association to get quota. In July 2022, when we went to sea, depending on where he went, he could have caught 50 kg of plaice in area VIIa or 4 tonnes of it in area VIId or VIIe. These are FAO fishing areas, often listed on fish packaging. In a month, Tom was also allowed 8 tonnes of monkfish and 2,666 kg of monkfish tails, no skate, no endangered rays, no spur dogs, and 10 tonnes of pollock. If his trawls brought up mackerel, he could land an amount that equalled 15 per cent of his catch.

Who can catch what fish and how much is decided in the UK by fishing quota. The UK, a country that has sold most of its national utilities to foreign buyers, allows fishing rights to be bought and leased. They are currency. Twenty-five companies now control two-thirds of the UK's fishing quota, and half of the quota is held by foreigners. Quotas are based on TAC, which is based on MSY. Clear? (Total allowable catch based on what scientists consider the maximum sustainable yield.) The UK at least is one of the few fishing authorities that sometimes sets quota according to scientific advice. Most countries and RFMOs routinely ignore the advised limits set by the International

Trawling

Council for the Exploration of the Sea and issue higher quotas. Because there is no one to stop them.

Now for the gutting. Tom wants me at the gutting table, a waist-high flat work surface with a raised lip that is attached to the port side. I know what he will ask, and I've been considering my answer. I am usually intrepid. I don't mind going down sewers or using foul latrines with no doors or climbing onto massive ships to spend many weeks with an unknown crew. I will do most things, but I will not gut living fish. He tries to persuade me. He says I should do it because it's part of the fishing experience. He says the camera crew did it. He may have called me a wimp. He may have been right.

Skipper Tom is efficient with his knife; the other Tom is learning. Grab a fish, slice it, throw it. Tom probably has a rhyme for this too. Gut it, chuck it in a bucket. 'That's it,' he says now and then to Tom, although the younger man is much slower. After a while I think to ask Tom whether the fishes are alive or dead. He looks surprised. 'Alive, I suppose.' I'm troubled both by the answer and that this articulate, smart man of the sea had not asked himself the question. He says, when the gutting is over, 'Right, that's your day observing a weapon of mass destruction.'

We think so differently about fish, he is starboard and I am port. At one point in the morning, dolphins appear and I am uplifted by their sleek beauty, their movement, how much they belong to this water, how they revel in it and in the great noisy machine they are zipping alongside for fun. They belong here. Tom is indifferent to them. 'Dolphins, you can get blasé about normal nature.' Instead he gets excited about tuna, which they have been seeing round here for the last five years or so. He marks on his electronic chart where he sees a shoal, just to remember. Maybe they are here because of less fishing pressure elsewhere, or because of warming waters. The simmering ocean is now pushing fishes into new areas.

'Tuna is a bit different,' says Tom, when I wonder why he is not more excited by dolphins. 'To me they are the pinnacle of evolution. Because they are built for speed and aggressive feeding, their fins will track into their skin so they're even more streamlined so they can sprint. They're actually slightly warmer than the sea temperature, they're a couple of degrees warmer so they can get oxygen into their haemoglobin.' I agree: tuna are majestic animals, and I assume his admiration is like that for a panther or a puma. But his next sentence is about how he has fished for tuna in Australia and the Bay of Biscay. 'There's nothing more magnificent than tuna. Especially when you catch them, their iridescent colours. Fantastic. Brilliant. I love to fish for tuna.' He thinks they are magnificent because they are worthy prey. I think they are enough in themselves. Starboard and port.

Someone who knows me well had worried that I would be uncomfortable on a trawler. They know that I don't like the smell of fishmongers. The trawler does not smell like a fishmonger. It smells like the sea. Fresh fish does not smell fishy. Fish counteract the saltiness in seawater by filling their cells with trimethylamine oxide (TMO). After they die, the TMO is converted to trimethylamine, which creates a 'fishy' smell. Fishiness is not confined to fish. Trimethylaminuria is a human condition that causes people to emit a fishy smell through sweat and other fluids because they can't convert trimethylamine into trimethylamine-N-oxide (TMAO) and choline as we are supposed to do. The condition can be worsened by sweating; eating fish, eggs, or beans; or having a period. There is no cure for what a paper on trimethylaminuria called 'A Socially Distressing Condition'.[17] The common name for trimethylaminuria is fish odour syndrome. A singer named Cassie Graves became briefly famous, telling the media about 'her daily battle to stop herself smelling of rotting fish'. Having fish odour syndrome would be bad enough, but sufferers can't smell it themselves: Graves only knew when her sister said she stank. 'My mum presumed she was just being the standard grumpy older sibling and ignored her. However gradually, the smell became so strong that no one could deny it.'[18] I'm surprised trimethylaminuria didn't appear in Dante's *Inferno* as one of his ingenious tortures. Heads on backward; submerged in boiling blood; having to rely on strangers to tell you that you smell foul.

A box full of handsome spotted fishes holds my attention. One dies upright, its facial features – that smile-shaped mouth – showing that these are sharks. The murgey, dogfish, or lesser spotted catfish, the Cornwall Seafood Guide informs me, 'are small sharks that live on and near the seabed. They have handsome spotted skins that are covered in tough sandpaper like skin teeth. It is very difficult to remove this skin and as a result dogfish is rarely eaten in Cornwall and many fishermen use it as bait. It is however edible and sometimes marketed as "sweet William".'[19] The Cornwall Fish Producers' Association is adept at such transformative marketing. When recently it changed the name of 'megrim' to 'Cornish sole' and 'spider crab' to 'Cornish king crab', sales of both increased.[20]

Tom treats me like a guest and a tourist on the boat, pointing things out, introducing our neighbours on the water. A small fishing boat painted a lurid colour. 'It's a really, really, really shitty green.' He knows the skipper. The boat belongs to a shellfish merchant. 'It had quite a big refit and suddenly appeared in Newlyn painted this horrible colour. You walk past it and you can almost taste it, it's that bad a violent colour.' When I ask if fishing keeps him fit, he points to another boat. It's a crabber. 'That's a very physical, demanding job, they're a lot fitter than me. I'll start sweating when you see me hauling.' He is not fit, he says. 'Trawling is not an overly demanding job physically.' He has done other types of fishing, but he likes bottom trawling best. As for the eco objections, he says beam trawling is worse. That really digs into the seabed because the net is attached to a beam that drags along it. 'A beam trawler is a very good way of catching flat fish but, depending on where you're working, it can be very destructive. If you work the hard way, it is damaging whatever they say. But when it's on fine open ground, that damage will right itself because the tide is moving the sediment.'

On the *Guardian*'s deck are some black lumps. They are bits of coal from old steamships that have sunk. You can trawl all sorts. The strangest thing that Captain Tom has trawled was a giant whale cartilage. Other fishing vessels have trawled war mines: 80 years later they still appear, and no government department knows how many there are still and when fishing gear might hit one.

It is a long day on the *Guardian*, and it ends at 7 p.m. with us landing the catch in front of Newlyn fish market. In the summer the days are even longer: departure is at 2.45 a.m. I climb up the ladder to the wharf and watch for a while. They'll be out again in eight hours, in the fishing hours. I'm going to bed.

Tom and I stay in touch. Usually Tom's messages ask me if I've read *Distant Water* yet, a book by the US journalist William Warner that Tom thinks is the most accurate portrayal of fishing life.[21] A month after I leave Newlyn, I read that the *Guardian* was involved in rescuing a yacht in trouble. The trimaran *Pir2* was on its way to France from Guadeloupe when it capsized five miles from Land's End. These are Tom's fishing grounds – whenever I go to a vessel-tracking website to find him, there he is, roughly the same distance and direction from shore – and from a mile away he saw the trimaran's mast disappear. There are no rules requiring boats to rescue other boats, just convention. Luckily, enough still do. Tom noticed from the boat's AIS that it had slowed dramatically, and he set off to help. By the time he reached the boat, a Navy helicopter on exercise had appeared and airlifted the skipper to safety.

He was sanguine about this when I messaged him. He would get paid salvage money, as the *Guardian* had got to the yacht before the lifeboat. 'First time in 35 years!' He continued to be helpful, putting me in touch with fish people who could get me to sea on other types of boats so that 'you get a balanced picture'. And if I want to go fishing on the *Guardian* again that's OK too. 'If you want to come back and vomit all over my boat again, that's fine with me. I promise not to laugh.'

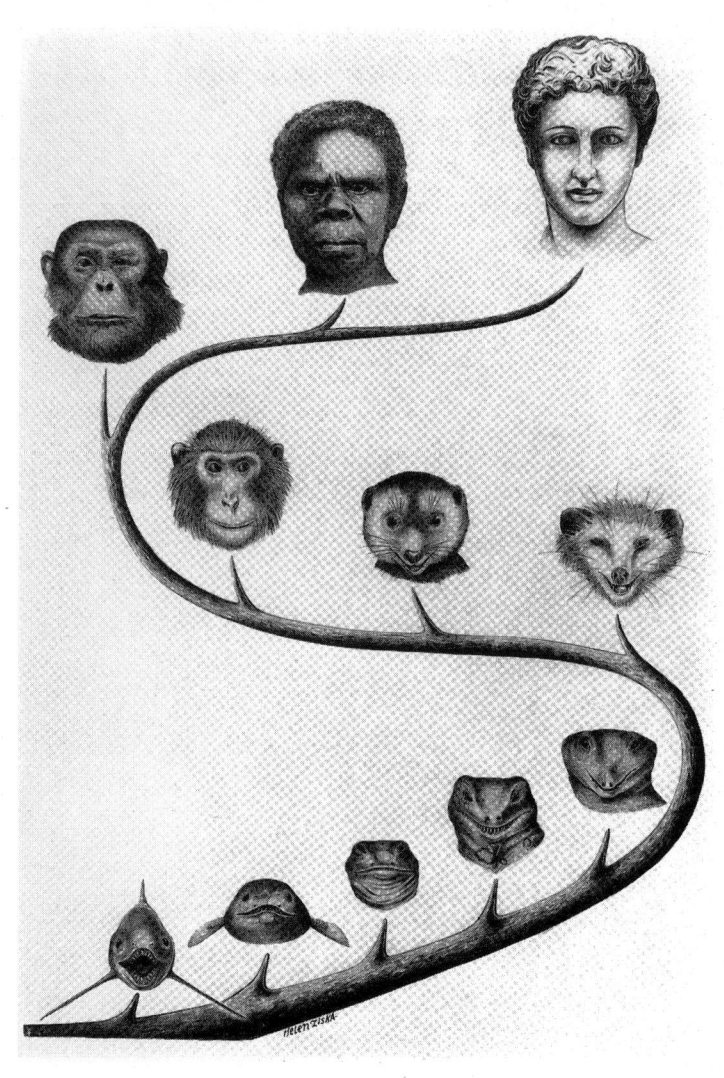

Our face, from fish to man

7

Fish Are Not Chips

Fish are friends, not food.

BRUCE THE SHARK, FINDING NEMO

A fish, says Culum Brown, is just 'another sort of potato'. He has been researching fish sentience for decades and still thinks that for most humans, fishes merit no more attention than a vegetable. Or fruit. In French, Italian, German, and Brazilian Portuguese, marine creatures are as inanimate as fruit: no matter how clever the lobster, it still becomes only *fruits de mer*, *frutti di mare*, or *Meeresfrüchte*, and it is assumed to have as little sentience as an apple.

I encountered Culum Brown first by reputation. He features on a Compassion in World Farming list of visionaries along with the primatologist Jane Goodall. What did you have to do to earn that? 'Basically put your head on a chopping block and wait for people to take a swing at you.' A podcast that featured him was called 'Dr Fish Feelings'.[1] His real title is professor in the School of Natural Sciences at Macquarie University in Sydney, but I also see him referred to as a biological ethicist, an eco-ethologist, an ichthyologist, a marine biologist, and 'a champion for fish'. The podcast introduction told listeners, 'Goldfish have feelings. Manta rays use facial expressions to greet their mates and some of them know when it's the weekend.' Humans and fish, it continued, 'are made essentially from the same box of Lego, just with a few tweaks'.

This is a surprisingly controversial idea even now. For decades, believing that fishes have feelings and intelligence was a lonely cause.

Plenty of people have no interest in taking an interest in the abilities of fishes, says Brown. 'Rec fishers, commercial fishers, aquaculture people. You name it.' Brown began by being interested in fishes for themselves. 'I just thought they were cool and I was interested in the things they could do.' His activism about improving fish welfare arose like a tide the more he understood about the animals and about how much we misuse them. 'Twenty years ago nobody talked about welfare. There were a few review papers by people who I knew, but they never really went anywhere. They didn't even make much inroads into the fish biology people, let alone into society. That's changed, particularly in the last five years. Even your average person starts to think about where their fish came from, whether it died well. It's been dragged along with the rest of the things we eat, free-range this and cage-free that.'

I tell him I haven't seen any Save the Goldfish campaigns recently. 'If you speak to NGOs, they admit they've done a pretty good job in terrestrial systems trying to get people to think about your chickens and cows and whatever else. They all admit they dropped the ball when it came to fish. They just went for the low-hanging fruit, go for the things that people naturally have empathy for.' Most people have encountered a sheep or a cow in nature or at least in industrialized agriculture, but a fish is usually dead on a plate.

Empathy often depends on similarity, and if humans can be convinced that fishes are smart and individuals with personalities, not heaps on a dock, they will defend them better.

I'm standing in front of the main fish tank at The Deep, an aquarium in Hull. I have accidentally chosen to visit on a school holiday, so the place is packed. The energy of dozens of excited children is charming but their noise is not. The noise has increased from vaguely interested to berserk. The shark is coming.

Sharks are fabulous and threatened, and *Jaws* has much to answer for. Only I'm not here for the shark. There, shimmying along, is a sleek little striped fish that no one is paying attention to. A wrasse. If the children knew what a wrasse could do, they'd surely be going berserk for that too.

I am bewitched by the wrasse. In fact, I am bewitched by wrasses because there are many varieties of this extraordinary animal. Cleaner, parrot, rainbow, and cuckoo. The beautiful six-line wrasse or the mystery wrasse, which features a fake eye on its tail to confuse predators; the leopard and the exquisite fairy. Wrasse are frequently chosen as aquarium fish because they are pretty and peaceful and simple. At least that's what the aquarium world thinks. Cognitive fish biologists know differently. In an essay on Machiavellian behaviour in fishes, wrasse are the stars (with a hat tip to the cichlids and groupers). The language is sober but it doesn't stop me saying every page or so, 'What? They do what?'[2]

They change sex. Cleaner wrasse are protogynous hermaphrodites and begin as female before switching to male. (In the other direction, they would be protandrous and that's a different kettle of fish.) Cleaner wrasse live in harems: a dominant male with several females. If there is no dominant male, the most dominant female becomes the male. The zoologists took eight pairs of cleaner wrasse from an Australian reef, put them in plastic bags filled with aerated seawater, and took them to a nearby testing lab. There, the fish were given various tests with impressive names: reversal learning task, detour task, audience effect task. The females were better at self-control, demonstrating this by learning not to swim directly into a plastic barrier that separated them from food, but going another way around. Males – who used to be females – were better at associating a colour or shape with a food reward. Between every two cognitive tasks, they got a day off. At the end of their service, they were returned to wherever they had been captured. On balance, I'd rather be a cleaner fish than a mouse when it comes to serving scientific knowledge. Being a lab mouse is usually a one-way trip to extinction with no holidays.

I'd also rather be a cleaner fish because they seem to have more fun than their name suggests. Cleaner fish are charmingly devious and not just by their sex-shifting. They get their name because they clean other fish. They are meant to target a client fish's ectoparasites. Actually wrasse prefer mucus so sometimes bite their clients to get a good mouthful.

I don't blame the cleaners for their transgressions. They must be exhausted. They have about 2,000 interactions a day, according to marine biologist Redouan Bshary, who has been counting. Luckily for the clients, there are also cleaner police: male cleaners spy on their harem's cleaning services and chase biting or mucus-stealing females. In fact, wrote Brown and Bshary, the presence of any onlooker means the transgressors 'are . . . more cooperative to current clients'.[3]

A 2018 study on cleaner wrasse used the mirror mark test, a common way to prove animal sentience.[4] If an animal shows that it can recognize its own image in a mirror, then it has self-awareness. When cleaner wrasse were marked and shown a mirror, they did not attack the image, supposedly because they knew they had nothing to fear from themselves. In 2024, researchers went further. Bluestreak cleaner wrasse were placed in a tank with a mirror.[5] Then the mirror was removed and a photograph of the wrasse with a mark on its face replaced it. They responded by trying to remove the mark on their face. When the image was of a different but similar fish, they attacked it as a rival. 'When fish were exposed to composite photographs, the self-face/unfamiliar body were not attacked, but photographs of unfamiliar face/self-body were attacked, demonstrating that cleaner fish with [mirror self-recognition] capacity recognize their own facial characteristics in photographs.'[6] In animal cognition this is huge. If a fish can distinguish its own face in an image, then it must have a mental image of itself. It has a 'self-face', as most humans do.

Fish abilities shouldn't surprise us. We know now that, as Culum Brown writes, 'fish show a wide array of sophisticated behaviours. For example, they have excellent long-term memories, develop complex traditions, show signs of Machiavellian intelligence, cooperate with and recognize one another and are even capable of tool use.'[7] So far 15 species of wrasse have been observed to use an 'anvil' technique: they grab their prey – an urchin or clam, usually – and smash it against a rock. It saves their teeth and gets them dinner.[8]

Fishes live in a complicated environment. They have to find food, escape predators, find mates, breed safely, avoid nets and hooks. Why would their abilities also not be complicated? Pike that have been hooked remember to steer clear of hooks – this is called 'hook shyness' rather than 'common bloody sense' – for more than a year. When rainbow fish learned to escape through a hole in a net in their aquarium, they could do it 12 months later even though they hadn't seen the net since. Brown sometimes sounds puzzled that people still question whether fish are sentient. 'Because they're vertebrates, it's pretty likely that sentience arose once in vertebrates and it happened early on.'[9] The physiology and anatomy of fishes are similar to other vertebrates including us. They are similar enough that the zebrafish is now a frequent choice for lab research. Researchers like its transparent embryos, its human-like cell structures, and its habit of laying 100–500 fast-growing eggs each week. (Mice breed more slowly and less fruitfully. Also you can't see through a mouse.)

The academic Kathy Hessler has been a pioneer of scientific studies into animal welfare. She set up the Aquatic Animals Law Unit at Clark University in Massachusetts, the first of its kind. In a lecture, she recalls encounters with animal researchers used to experimenting on mice and rats.

'We're doing replacement, we're replacing animals.'

'What are you replacing them with?'

'Zebrafish.'

'That's an animal.'[10]

What a fundamental irony, Hessler says, that fishes are considered close enough to humans to be useful scientifically but not close enough to be protected in law. In a lab, fishes are our kin. In a fishing trawl, they are not.

Research has advanced late but fast, and technology helps: now scientists can track ocean animals for years. So we know now that, like rays, says Brown, sharks can have 'complex despotic social societies, the kind you see with primates'. Rays are clever enough to learn when recreational fishermen clean their catches and to wait in anticipation. Brown's lab has also shown that rays like to hang out with other rays, but not just anyone. They choose their mates, as we do. 'All of the work we've been doing on sharks and rays has blown out of the water this kind of concept that they're kind of boring and they're not really doing anything interesting. They actually are. Most people think of sharks and rays as being these asocial aggressive things, but far from it. They're far more complicated than we ever thought.'

Still, sharks' undeserved reputation is useful in getting funding to study them. 'At least if I write a grant about sharks,' says Brown, 'people go, ooh, sharks, they're scary and they're kind of interested. But if you're just writing about fishes, no one really cares.'

If fishes have feelings, will we treat them better? We now know how intelligent and amazing the octopus is (actually a cephalopod not a fish), and humans eat ten times more octopus now than in 1950.[11] Plenty of people love to see lambs in springtime and have them for lunch. We could not live in dense cities without what American sociologist Erving Goffman called 'civil inattention'. This is the ability to ignore the fact that we are surrounded by billions of people. We are good at ignoring inconvenient truths. For Culum Brown, 'the ramifications for such animal welfare legislation, should it be applied to fishes, [are] perhaps too daunting to consider. Certainly, the fishing industry, including aquaculture and recreational angling, would have to drastically alter its approach.'[12] Elsewhere his language was more Australian. 'One of the problems with announcing that these animals need protection is that the industry would go berserk.'[13] Jonathan Balcombe responded to a review of *What a Fish Knows*, his book on fish sentience, by saying that the reviewer 'believes that I am trying to humanize fishes, when in fact I am seeking to inject some more humanity into humans'.

Ethicists and philosophers talk of humans having a moral circle. Inside the circle are creatures that matter enough for humans to care about their welfare. In most laws, fishes do not have morally relevant interests. They cannot be allowed to have them, despite all the solid research proving they are smart and playful and sensitive, because the fishing industry would be in trouble if they did. 'Efforts to improve fish welfare,' wrote the authors of the paper 'Animal Minds, Social Change, and the Future of Fisheries', 'are generally secondary to the primary goal of exploiting the fishery for food or profit.'[14]

When PETA put an advertisement in a UK bus shelter showing a fishmonger holding up a fish that turned into a dead cat (using a hologram), a passerby, asked for his views, pronounced the ad 'a bit sick'. 'It's a cat. You don't eat a cat.'[15] PETA said the ad demonstrated 'speciesism'. When it comes to suffering, 'there is no difference between a fish, a cat or any other animal'.[16] The response from the National Federation of Fishermen's Organisations was expected. 'The people of the UK have enjoyed eating fish for thousands of years,' said Mike Cohen. 'To suggest that fish feel pain in the same way as mammals is, at best, highly misleading.'[17]

Much debate about fish sentience is still – amazingly – about whether fishes feel pain. Let me help you out. Of course they do. We are far from the days of Peter Singer when he wrote in *Animal Liberation* that 'with creatures like oysters, doubts about a capacity for pain are considerable; and in the first edition of this book I suggested that somewhere between shrimp and an oyster seems as good a place to draw a line as any'.[18] Happily, Peter Singer is also far from those views: a 2010 op-ed he wrote was entitled 'If Fish Could Scream'.[19]

'Can fish feel pain?' wrote William Strange in *The Gentleman's Magazine*. Then he answered himself. 'Undoubtedly they can. A vertebrate animal, endowed with a complete nervous system; – nerves, spinal cord, and brain, and yet incapable of experiencing pleasurable or painful sensations, would be an anomaly inconceivable by the mind of a Cuvier or a Huxley.' He wrote this in 1870 and had been triggered to respond to a pamphlet by Mr Cholmondley-Pennell with the same title, which declared that 'pain, in the sense in which humans are conscious of it, is unknown to fish organisation'.[20]

A famous paper by Lynne Sneddon featured 25 rainbow trout bought from a commercial fish supplier. They were placed in a tank then some had acetic acid injected into their lips. Others were injected with saline, some with morphine, and some with acid and then morphine. The acid-only fishes showed increased opercular beat rate (gill movement), a standard measure of pain response. The acid-only trout and the acid–morphine trout showed rocking behaviour, and the acid-only trout rubbed their lips against the side of the tank. Scratching an itch. The acid-only fishes also took much longer to resume eating. So would I if you had injected acid into my lips.

Even with research like this, plenty of people will not accept that fishes have feelings. Their usual argument involves the fish brain. Because fish lack a neocortex like humans, wrote neuroscientist Brian Key, 'they cannot feel'.[21] 'But that,' wrote the ecologist Carl Safina, 'is like saying that because we travel using legs, then fish, who have no legs, cannot travel.'[22] The neocortex, unique to mammals, is thought to be the most evolved part of the brain. 'Birds,' writes Jonathan Balcombe, 'lack a neocortex, yet the evidence for consciousness in birds is virtually universally accepted.'[23] Birds use a 'paleocortex', an older part of the brain, and still manage to be clever; why not fishes?

Another assumption is that fishes are dumb because their brains are small. But that's us being dumb, as Sy Montgomery wrote in an essay on the octopus mind. 'One yardstick scientists use is brain size, since humans have big brains. But size doesn't always match smarts. As is well known in electronics, anything can be miniaturized. Small brain size was the evidence once used to argue that birds were stupid – before some birds were proven intelligent enough to compose music, invent dance steps, ask questions, and do math.'[24]

Last year, German researchers established that cichlids and stingrays can add and subtract.[25] When human adults are disoriented in an oblong-shaped room that has one painted wall and are asked to find an object, they can use the painted wall to orient themselves. This is called a featural cue, and children can't understand it until they are six years old. Fishes can. Rock-pool-dwelling gobies know where their home pool is even when they have been displaced by 30 metres. Even after being removed from their home pools for 40 days, the gobies could still remember the location of surrounding pools.[26] This is called a cognitive map, and gobies the size of my hand have one as good as mine. (Given how easily I can get lost in the hills, it is probably superior.) It is, writes Brown, 'an astonishing ability'.

Recently a team from Israel decided to see if goldfish could drive, and the world should thank them for that inclination. The team devised a Fish Operated Vehicle (FOV), which they describe as 'a chassis measuring 40 × 40 × 19 cm that housed the platform on which the water tank was placed'. The FOV moved according to the goldfish's movement, and the goldfish was meant to direct the FOV to a target. No bother for the goldfish, those derided pets: They 'indeed were able to operate the vehicle, explore the new environment, and reach the target regardless of the starting point, all while avoiding dead-ends and correcting location inaccuracies'.[27] (Goldfish can also remember things for months, not three seconds.)[28] If fishes are fed at different ends of the aquarium each day, they can learn to anticipate where they will be fed. Wild rainbow fish learned to hang around at the correct end after 14 attempts. When rats have to associate a tone with food, they learn it after 40 tries.[29]

In my notes: 'Smarter than toddlers. Quicker than rats. And they can drive.'

A recent review of the continuing debate into whether fish feel pain recently lamented the 'research inertia' and lack of new exciting papers. Inertia is not just in research. Even with all the new research about fishes, we still kill trillions each year and do not consider their welfare. In a maelstrom in the Arctic Circle, some Norwegian divers are trying to change that, with creativity and courage.

Pernille was with Thærie in a cosy cave but Thærie disappeared, so Pernille took up with Konrad and they made babies. They were content for a season then an aggressive male intruded, and Konrad was scared off. The bully moved in with Pernille, but she threw him out. She moved out of the cave and has not been seen since.

This underwater soap opera concerns ugly-charming wolffish. They are charming because of their grace in the water, their large bodies swishing as they move. They are ugly, because have you seen their faces? These wolffish live in the majestic and unique surroundings of Saltstraumen in northern Norway, a strait like no other in the world. The water flowing between inland Skjerstadfjord and outer Saltenfjord, combined with the depths and the width of the channel, have made the world's strongest tidal current. Every second, 3 metres of water hurtles through the 150-metre passage, sometimes at speeds of 20 knots. It produces the biggest maelstrom on the planet.

In 2013, Saltstraumen became a marine protected area (MPA). On paper, Norway comes under 'could do better' when it comes to MPAs. Less than one per cent of its waters is protected, although that is better than most countries manage. Norway's rate of assessing its MPAs is, according to the Marine Protection Atlas, zero.

For Borghild Viem and Fredric Ihrsen, who run a diving centre on the banks of Saltstraumen, the protection of their maelstrom is fiction. The Visit Bodø tourism site extols Saltstraumen's rich marine life. 'The abundant stocks of cod, saithe, halibut, monkfish and wolffish make it an Eldorado for anglers and seabirds such as eider ducks and white-tailed eagles.' The only part of this rich marine life that is actually protected is kelp. In this protected area, fishing is allowed; underwater hunting is allowed; the discarding and abandonment of thousands of fish-strangling lines and hooks are allowed. (The gathering up of those hooks and lines though requires a permit.)

I had met Borghild and Fredric the previous day at a university workshop that launched a two-year EU project to establish where best to site MPAs. Much of the day involved complex and technical science: Borghild and Fredric occasionally looked as mystified as I felt, and they were divers, the best underwater monitors we have, so I was drawn to them on both counts.

Fish Are Not Chips

On the day I visit them at Saltstraumen, the weather was sideways snow and cold. We are just above the Arctic Circle here, but this weather has surprised locals: they thought they were done with winter. It took three buses and 90 minutes to get here, so I am relieved that Fredric has taken pity on me in this weather and collects me from the bus stop. That way he can point out the old wooden house that is the Nord NE diving centre, which sits sedately on the shore, grey forbidding water beyond. Inside, Fredric serves me first a pair of fleece-lined house slippers then good dark chocolate and hot coffee. He offers a Scandinavian choice of mugs. 'Pippi Longstocking or the Moomin one?' He explains that Borghild is underwater, 'out there somewhere', so for now it is up to him to tell the story of the Saltstraumen wolffish, which may be the salvation of this troubled MPA.

A young blond man is in the kitchen working hard on a laptop. He doesn't look as if he is doing anything historic, but he is because he is installing a camera. For the first time, underwater cameras are being installed in Borghild and Fredric's diving area. It is a project of Nord University, sanctioned by the state government.

Wolffish – also known as catfish, devil fish, and sea wolves – are benthic bottom dwellers, mostly: they only move to eat, and although their faces are ones only a mother could love, their colours, shades of shining marine blues, are beautiful. And their behaviour is mesmerizing, because once the female has laid eggs, the father guards them in his cave for 5 or 6 months, not eating and not moving. Its Latin name, *Anarhichas lupus*, takes the 'lupus' part (wolf) from the fish's impressive front teeth. Wolffish are important predators because those teeth let them eat clams and the ubiquitous sea urchin, which I saw all over the underwater landscape of the Bay of Dakar because there are too few fishes left to eat them.

Wolffish are also tasty and in Saltstraumen underwater hunters come to fish them, usually stabbing them to death because the animals stay put and are large. Easy targets. Borghild and Fredric decided to launch a defence. They didn't like the hunters, they didn't like the litter they abandoned in the water, and they did like the wolffish. Their method was to create a wolffish soap opera. 'We want to personify the MPA,' says Fredric, 'so we make stories. The wolffish has been the red thread through the stories.' Their first wolffish was named Thærie, after a Henrik Ibsen poem about a heroic maritime pilot named Terje Vigen (Thærie is an older version of the name Terje).

Thærie was the start of their wolffish education. 'We suddenly met this, wow, it's a wolffish with eggs. Then we started to follow him.' By simple observation over time, Borghild and Fredric discovered a fact that had eluded marine biologists: wolffish mate for life, or they try to. 'We saw that it was the same lady who came back, and then the next summer the same lady again.' They became experts on the fishes' life-cycle. Lovemaking in late autumn; in November the eggs were laid, and then the 'lady' leaves but the humans don't know where she goes. Thærie would stay put, barely moving, wrapped around the 5,000 or so eggs, until late April. 'When the eggs are hatched,' says Borghild, 'I guess Thærie is quite bored just lying in the cave, so he usually leaves and the cave is empty. In June, he goes to and fro a bit, fixing the house, and in July the lady comes back. Year after year after year.'

Divers are meant to leave underwater wildlife in peace. You don't touch, you don't disturb. Yet I come to think of Borghild and Fredric as being the most hands-on of environmental protectors. They understand that people will protect what they have come to love and that often they love what they can identify with, so after Thærie, they began to name their fish couples after Saltstraumen villagers from history. Pernille was a local midwife. 'A very important person in the village, because they had no doctors, so she did healing too. Konrad was a one-legged sailor who was custodian of the lighthouse. Parelius and Ida, another couple, were named after the village ferryman and his wife.' (A bridge was built in 1978 to reduce the danger of the crossing.)

Borghild is convinced from her observations that there is wolffish true love. 'He was lying there waiting, waiting, a lot of beautiful wolffish were swimming around, he's no, no, no and then one day he was literally dragging one with his fins and the next day she was in the cave. She is quite beautiful and big and he is a bit shy.' This was Pernille. 'They stayed together four years.'

There were troubles in those four years, and after, and some of the biggest troubles were caused by Borghild and Fredric. Captivated by the wolffish romances, they dived and took pictures of freshly laid eggs. 'After a month we learned that that is not a good idea, to take pictures of freshly laid eggs. A month later, the eggs were gone. And Thærie was gone, the cave was gone. We were like, what's happening?' Aquaculture scientists enlightened them. Enlightenment was the problem: freshly laid eggs are sensitive to bright lights. A camera flash is a bright light, and the light destroyed them. Borghild was horrified by what they had done. 'We had maybe destroyed a whole generation. Murderers. We were almost crying, it was really hard.'

Nature forgave them, because the following year, eggs were laid again. And now Borghild and Fredric don't permit any photography from November to February. 'We think that's why they lay the eggs in November, as it is dark.' The cold is no hindrance to wolffish happiness: wolffish make their own antifreeze that prevents their blood from icing. Pernille and Thærie's love story ended in 2021. 'Suddenly all the wolffish were gone,' says Borghild. There were underwater hunters in the area, and for a week all the wolffish caves were deserted.

In the UK, the fish is called Scarborough woof, and sometimes used by fish and chip shops. In the United States, the Atlantic wolffish is a species of concern and Canada calls it a 'species at risk'.[30] In the UK, using as a measure how many are caught as by-catch, wolffish have declined by 96 per cent since 1889,[31] a remarkable decline for a fish that is not generally caught for food. There are no formal wolffish fisheries but their bottom-dwelling habits make them vulnerable to trawling. Wolffish are vulnerable not just because they don't move but also because they mature relatively late and only breed after the age of seven or so. This means stock that is depleted takes longer than other populations to recover. In a piece headlined 'WE'RE ACCIDENTALLY DRIVING THIS EXTREMELY UGLY FISH TO EXTINCTION', Tom Murray wrote that, 'the wolffish therefore faces the worst of all worlds: It is being hunted to extinction by accident, and – unlike the tiger or the panda or the snow leopard – isn't cute enough to have attracted any humans passionate enough to save it.'[32]

In her presentation, Borghild has come to what she calls 'the bad pictures'. Kelp forests strewn with fishing lines. A fish trapped by a lost hook and stuck to the sea bottom, wrapped in and entangled by fishing line. It is a horrible sight. Of course, Borghild freed him, although she thought it was hopeless. 'We thought, OK maybe he's dying but we'll give him a last swim, and the next day Fredric was diving in the same place, and the same fish, with the same scars, came up to him. "Look at me! Look at me!"'

The cameras are limited in where they can be placed by the length of cable available and by the terrain. One camera will look down on the wolffish caves; the other is in a location chosen because of 'a combination of how far we can reach, the most direct, and . . . an area [that is] most interesting for the halibut and cod. Also it's quite protected, with high walls.'

Watching this footage is not permitted. The state government has forbidden the university from sharing any images of the camera. Borghild and Fredric don't know why. The university professors don't know why. I don't know why: this footage can only help. Imagery is the best weapon of protection: photos of fish entangled in fishing lines and hooks feature heavily on the Facebook group, where the soap opera continues and will do until the hunters have gone, the underwater is cleared of fishing litter, and the wolffish are left in peace.

No permit is required to hunt or fish in this MPA. The only thing preventing industrial fishing is the force of the water. The richness of the underwater world attracts fishermen and hunters, and there is nothing to be done about it, even though the establishing law states that 'animal life with a connection to the seabed shall be protected from harm and destruction' and 'littering is prohibited'.[33] The same law forbids the harvesting of any vegetation. Borghild and Fredric are right: their beloved wolffish may belong on this seabed, but a plant is more protected than they are.

Fish Are Not Chips

Borghild is vivacious and cheery but also gloomy. 'More and more people are beginning to discover this paradise underwater. Most come from Denmark but there are also other Europeans. We have to stop this kind of fishing; we have to establish just a few places to fish and the government should pay to clean up the gear every year. But before, the pollution was in just a few areas and now it's all over.'

Borghild can only hope that her wolffish soap opera makes locals want to protect their MPA as much as she does. As for the fishing tourists, all she can do is 'talk, talk, talk': to the local hotel, to the people who rent fishing boats. 'They are trying to make people fish outside the current, not always fish in the current in the protected area. But we want them to say, "Don't fish." Instead they say, "Don't fish more than one or two wolffish inside the current."' Borghild sounds tired. It is hard work, but it must be done. 'Pressure, pressure, just tell people what's happening. We need to take care of this place. It's so beautiful and so special.'

Hopefully enough people will be interested in the love life of Pernille and Konrad, or Konrad's replacement, that they will pay attention to the whole environment of Saltstraumen and wonder about the litter and the killing. Hopefully they will check in now and then on the Facebook page. As I wrote this in autumn, Borghild emailed with news. The Ministry of Trade, Industry and Fisheries had just made an announcement. 'The wolffish in Saltstraumen marine park is now protected! So no one is allowed to catch wolffish any more. Not underwater hunters, not fishermen.' Otters though are exempt and can still hunt wolffish. 'I guess they are not included since they do not read or have internet. But they are not the problem, humans are.' If fishermen hook a wolffish accidentally – I'm not sure how – they have to let it go free unless it is too badly wounded. Borghild sounds delighted. I am delighted. And the wolffish, deep in their caves, minding their own business? 'I guess our friends deep down there can relax a bit more.'

© Photograph by kind permission of the Estate of Petty Officer Harold Dodgson BEM

The crew of Hull fishing trawler Kingston Cyanite

8

A Tin-Can Navy

No grave but the sea

TOWER HILL MEMORIAL, LONDON

In images, it is terrifying. Arcturus, a star, is far from being twinkly and sweet. It is a ball of red fire, because it is a red giant star, just under 37 light-years from earth. If light-years were miles, I could get to Arcturus in seven hours or so, on a good day and with a fair wind. Arcturus is visible from our land, so it has always been essential to the men and women who went to sea. Romans navigated with its light; so did early navigators in Polynesia. It is one and a half times brighter than the sun and the fourth brightest star visible in our sky.

Arcturus then is a good name for a vessel. There is *Arcturus* the bulk cargo ship flagged by Liberia, *Arcturus* the Norwegian standby safety vessel, and many leisure yachts. But I'm not looking for them. I seek *Arcturus*, Fishing, PH7979, currently 'engaged in fishing' in the English Channel, shipping forecast region Wight. This *Arcturus* is a scalloper. And this *Arcturus* is heroic.

The headline in *Fishing News*, whose readers know what a scalloper is, read 'SCALLOPER CREW SAVE 31 LIVES IN CHANNEL RESCUE'.[1] My subheading: 'When they didn't have to'. It happened on 14 December 2022, in midwinter on the English Channel. As usual, many people were heading in rubber dinghies from the French coast to the British one, because they had paid traffickers for the crossing and because they believed life would be better there. In December, the UK government recorded 1,744 'migrants detected' in 36 boats. 'Please note,' the page reads, 'that the data for 14 December 2022 does not include the four individuals who were confirmed dead following the serious incident.'[2]

These small dinghies and boats do not show on vessel-tracking websites. A fisherman can see a dinghy on radar, but he would have to be looking and the Channel is a busy place with lots to draw the attention. On 14 December, the scalloper *Arcturus* was 30 miles west of Dover when the crew heard people screaming. In December, the water temperature was probably about 10 degrees Celsius by day. At night, it would have dropped to under six. Many efforts have been made to calculate a definitive graph for how long a human body can survive in cold water at certain temperatures. There are too many variables: how well the human is dressed, how fit or fat they are. A fit and healthy human with a buoyancy device may survive an hour or two. Terror and panic together with cold water shock can reduce survival to minutes. What is certain is that travelling for many hours on an open boat in December, probably inadequately dressed and highly stressed, is the worst preparation for being thrown into a freezing ocean.

At 2.53 a.m., Utopia 56, a French NGO that works in the migrant camps of northern France, received a call via WhatsApp. 'Hello brother. We are in a boat and we have a problem. Please help. We have children and a family in a boat. Water is coming in but we don't have anything for it, for feeling safety. Please help me, please, please, please. We are in the water. We have a family.'[3]

At 2.57 a.m., Utopia 56 alerted the French Coastguard. The UK Coastguard says it was alerted at 3.05 a.m. Five minutes earlier, a crew member on *Arcturus* had heard screaming and rushed to wake the skipper. He said, 'There are migrants alongside the boat.' Raymond Strachan, captain of the *Arcturus*, crew of six, dashed outside and saw a man clinging to his boat. 'One guy was hanging off my wire. I thought at first it was just him, and once I got my fishing gear up – which took about three minutes – I stopped my boat and ran outside, and along the port side there were five of them hanging off the side of my boat.'

A Tin-Can Navy

A crew member took a video and it looks like hell. 'It was like something out of a Second World War movie,' said Strachan. 'People in the water everywhere, screaming.' For two hours after calling the Coastguard for help, the crew pulled people from the water. They were concentrating on the port side of the boat. Once they had everyone safe whom they could see, they set off to harbour.

'One of the crew shouted, "Whoa, there's a rope on the starboard side of the boat." The next thing I knew this rope was attached to a dead body. We had been concentrating on the port side and this one person had swum to the starboard side, tied a rope onto my fishing gear, and tied it around his wrist to keep himself alongside the boat. When I put my boat into gear his body floated up.'[4]

Three other people died before they could be rescued. The lucky ones on *Arcturus* were given a lukewarm shower and the crew's spare clothes and were put to bed in the crew accommodation, tucked up in warm quilts to get their body temperature back to a survivable level. The men of the *Arcturus* rescued 31 people and they didn't have to. There is no obligation on a fishing boat to pull 'irregular migrants' from the sea, but there is a moral code. 'When people's lives are in peril, you have to save them,' Strachan later said. 'You cannot leave people in the water.'

They had had practice at this. In January of the same year, the scalloper had come to the assistance of three small migrant boats, including one with five people in. 'They were paddling across the busy shipping lane and not a single one of them had a life jacket.'[5] In 2022, 45,746 people crossed the Channel for a better life or safety or both.[6]

Of the 31 people Strachan saved, many were children. Three were Afghans, two aged 12 and one aged 13. 'I was quite emotional to hear this had happened to a young boy,' said Strachan. 'I cannot imagine being 12 and being in that water. When I was that age, I was in the comfort of my own home, looked after by my family, I was making go-carts out of old crates, not being fished out of the sea.'

By any measure, the *Arcturus* crew were heroic. The next day, I read a press statement from the RNLI on social media. It began well. 'Our thoughts are with the families and loved ones of those who sadly lost their lives in yesterday's tragic incident in the English Channel.'

Then: 'In the very early hours of 14 December, HM Coastguard tasked RNLI volunteer crews from Dungeness, Ramsgate and Dover to an incident in which a small, inflatable boat, crossing the Channel got into difficulties. On arrival, crews were faced with the harrowing situation of a number of people distressed and in the water. Volunteers carried out the rescue of a number of people and returned them to Dover before subsequently re-tasking to search for more casualties.' Eventually, the RNLI thought to mention 'the fishing vessels whose actions on scene clearly saved the lives of many casualties'.[7]

It was 31 lives saved, and it was one fishing vessel, and really it should have been named and not shoved into the last line of a press release. Fishing vessels routinely help each other at sea even though there is only expectation, not obligation. 'Other seafarers often regarded trawlers as the "lifeboats of the sea",' wrote Robb Robinson in his history of trawling, 'for the trawlermen's rough and dangerous life made them fearless and skilful rescuers.' Not every time. Once a smack – a sailboat fishing vessel – was heading home towards nightfall, 'when those on watch spotted some mysterious figures dancing up and down on the water. Closer investigation revealed them to be six life belted corpses in perpendicular position and this ghastly sight so terrified the crew that they crowded on sail and put as much distance between themselves and the spot as possible. They were still in a state of deep shock when they tied up at Lowestoft.'[8]

In a perilous job that requires fortitude, fishermen are still often courageous beyond necessity. They are rarely praised for this. And the last time fishing vessels were properly saluted for their bravery was when they weren't just in danger, they volunteered for it.

They looked like armoured tin cans. They were known as 'minor vessels'. They were fishing trawlers and drifters, more used to hauling nets and catching fish, but with the First World War they became something else. They became His Majesty's Trawler and His Majesty's Drifter, fitted with 12-pounder guns and given racks of depth charges.

In 1907, Lord Beresford, Admiral of the Channel fleet, had a good idea.[9] If trawlers can catch fish, why can't they catch mines or submarines? Beresford saw the potential in the hardiness of both the boats and the men who crewed them. He got a couple of trawlers and fitted them out as experimental minesweepers. The fishing gear was switched for minesweeping gear, a cutting cable strung between two trawlers that worked together as a pair. The cable would slice through the mooring lines that held mines in place underwater. The trawlers were a success and seven years later when war broke out, the Royal Naval Reserve (Trawler Section) was ready. By 1915, there were 238 minesweeping trawlers operating in British waters. Not all RNR men were fishermen, but many were. They had special treatment: they were allowed to wear moustaches without a beard, unlike naval officers.[10] 'This privilege is guarded very closely by the RNR,' wrote an *Evening News* reporter, 'who, I believe, grow a moustache as a matter of principle on being transferred.'

In 1916, the journalist J. W. B. Chapman spent a week with the 'minor vessels', and his subeditor decided to headline his story 'THE LIGHTER SIDE OF PATROL DUTIES', which probably was not popular with men who were as likely to be torpedoed, strafed, and killed as anyone on a naval vessel.[11] The fighting fisherman, Chapman decided, was 'full of quip, joke, and crank', and his admiration for these fishermen-turned-fighters is clear and emphatic. 'There are upwards of 40,000 of these men serving in various capacities, and a gratifying but astounding fact is that this vast complement of beings, who are as tough as steel and know the North Sea from A to Z, sprung up . . . in one night. Just as a man may walk into a tailor's shop and step outside wearing a ready-made suit, so an "Order in Council" said, "Let there be an Auxiliary Navy" and the ships and men were there the next morning.'

The men were there with boats named *Mary Ann* and *Bonny Kate*, and *Meadow-Sweet* and *Marguerite*, but that would not do: soon these sweet-sounding fishing boats were HMT *Iron Duke* and *Victory*. (In the

Second World War, a fighting trawler named *Formidable* was delivered mail meant for a nearby merchant vessel with the same name. The merchant vessel was bigger and wanted to keep its name. The fishing trawler was forced into a rename. Once *Formidable*, it became HMT *Fidget*.[12])

Chapman's sailing companions are a bunch of characters, from Marco the Sparks (radio operator) who was 'always spruce', so that 'if the ship were going down he would have brushed his hair', to a deckhand who claimed he had never seen a man drowned. 'But I've sailed with hundreds who ought to have been.' The chief gunner 'seemed to be married to the gun', and 'his hobby was washing his clothes, and giving the glad eye to the sea gulls'. Chapman's conclusion after eight days and nights at sea is: 'These fishermen are a magnificent lot, full of play as a kitten, often sharp of speech, yet, if necessary they would undergo any risk to save a shipmate.'

Their job was to keep the sea lanes open and the fishing grounds operational. This meant minesweeping but also dealing with U-boats and aerial bombardment. Danger from the sea and sky and from their own team. If they were in a convoy and their boat sank after a U-boat attack, men who escaped were often then hit by depth charges launched by friendly vessels at the diving U-boat.

And what of the fishing vessels that were meant to keep fishing? Some had elderly crew because all the younger men were by now enlisted. And all were targets. On 25 September 1916, nearly the whole Scarborough trawler fleet was sunk by a U-boat attack. The *Tarantula*, the *Otter Hound*, the *Game Cock*, the *Sunshine*, the *Nil Desperandum*, and the *Quebec*. The skipper of the *Otter Hound* had been on the *Dalhousie* a few weeks earlier when that was sunk. No fishermen died during the attack; the U-boat stopped a Norwegian steamer and asked her to take the captured fishermen on board. At this stage, the fishermen were not targets, only the boats.[13]

Fishing vessels were supposed to be protected under the Hague agreements in both World Wars. In 1940, a German court interpreted this in an interesting fashion, deciding that only fishing vessels 'belonging to poorer people' should be exempt from attack. This, the court said, was 'an exemption based on humanity'.[14] Trawlers were prey; smaller day boats were spared, even though the crew may get the same take-home pay. Tom would have got hit. Rex's catamaran would have been sunk and his salmon boat spared.

I become fascinated by a skipper named Thomas Crisp because of two photographs of him. In the first, he stands with his family and is dressed in a fishing sweater and a fishing cap set at an angle. He has a moustache and looks like the trawler skipper he is. The other photograph is him in RNR uniform, now clean-shaven, wearing a military cap. In 1917, Thomas Crisp was captain of a Q-ship. These were small fishing boats such as coasters or fishing smacks which had guns concealed on board. Their role was to lure U-boats and then wreck them with weaponry the U-boats would not expect to find on a small vessel. By 1917, the Germans had realized the ruse, but even so, Crisp's boat *I'll Try* sank a submarine in January, which explains the Distinguished Service Medal he is wearing in his RNR portrait. By August, the *I'll Try* had become the *Nelson*, and it was fishing off Lincolnshire. They saw a U-boat approaching, which shelled the *Nelson*. One shell hit Tom Crisp directly, blowing away half his body. Somehow he could still speak, and told the crew to launch a carrier pigeon with the message '*Nelson* being attacked by submarine. Skipper killed. Jim Howe Bank. Send assistance at once.' His son, also Thomas, was in the crew, and Thomas senior told his son to escape and to throw him overboard. The lad refused, then his dad died in his arms. The Germans had failed to shoot down the pigeon – a male named Red Cock – and it carried its message safely to Lowestoft, which sent a rescue ship for the survivors. Tom Crisp was posthumously awarded one of only two Victoria Crosses given to fighting fishermen.[15]

Tom Crisp is the image of what a whole industry had to do: to transform into something else, a fishing vessel but armed with guns and depth charges and de-mining equipment, or a fisherman who became a fighter, who remained a fishermen even in naval uniform.

By the end of the war, 672 fishing vessels had been lost, and 416 fishermen.[16] The RNR had performed so brilliantly in the First World War, of course it was resurrected in the Second. This time, skippers and crews were sent to Sparrow's Nest, a vaudeville theatre in Lowestoft, for some hasty training (three weeks was usual, though some oral history accounts talk of three days). Then they were sent directly to the wartime sea, sometimes in wooden boats. They were officially now named the Royal Naval Patrol Service (RNPS) but soon gathered nicknames: Churchill's Pirates. The Lilliput Navy. The name that stuck best was Harry Tate's Navy, after a well-known comedian who traded on his inability to deal with modern gadgets and inventions. His most famous sketch involved a car that collapsed around him. He was the byword for clumsiness and feather brains, for gormlessness and ineptitude, but also for the courage of the small man floundering in the face of big change he does not understand. It was meant to be a slur, spread by 'real' naval personnel – what use were these shabby fishing vessels at war? – so the fighting fishing vessels adopted it as a mark of pride.

Paul Lund, who recorded his RNPS experiences in a book written with Harry Ludlam, joined up as soon as he could. 'On the third day we were made up into crews, given £5 each to post to our wives and sent off by bus to Hull.' Lund was not a fisherman but he learned to think highly of them. 'These Patrol Service fishermen were tough men. Out trawling they thought nothing of working 60 to 70 hours on deck without sleep; they could carry on gutting and washing fish automatically, and fall asleep at meals eaten while on the job. Like the man who was given his breakfast of fish and fell asleep over it; his mates took it away and replaced it with a plate of fishbones. When the fisherman woke up he carried on with his work immediately, thinking he had eaten his meal.' The fighting trawlers flew a white ensign, not the usual red one of the 'proper' Navy, but a white ensign cannot wave away fishermen's beliefs that keep them safe.

'We had a black cat,' an unnamed telegraphist remembered, 'which all our fishermen crew thought was very lucky, and once when we were due to sail they all stepped on to the jetty at Harwich and refused to go to sea, because the cat was missing. After a thorough search of the ship we found the cat in the provision locker lying under a sack of potatoes, nearly flattened yet otherwise unharmed. And off we went to sea. We always wore our lifebelts at sea and the cat was the only one aboard without one. So one of the seamen got a Durex, blew it up and tied it round the cat's neck. It looked so funny, but we got used to it, and the cat went around like that for months.' (Condoms were also useful for storing money, as the rubber could survive a sinking. And for other activities.)[17]

Seaman Robert Muir, a clerk, learned fast about the fishing life. He was billeted to a small drifter with a toilet that never worked, but the fishermen were used to that. 'In peacetime they reckoned that when the weather was bad they just performed on the fish, and they solemnly advised me never to eat fish.' Even when they passed a minesweeping trawler and swapped tobacco for fish, no fisherman would touch it.

At first the men were relatively young. The original appeal was for fishermen with a year's deep-sea fishing experience, age limits 18–45. Men with 'a satisfactory general experience' would also probably be accepted. The requisitioned trawlers and drifters were as old as their crews. Some had been at sea for 30 years or more. And now they were meant to defend the nation from underwater and from the air, from machinery and vessels several times their size, with men more used to fighting struggling fish not Nazis, with senior officers who insisted on being called Skipper, who had nothing of the Navy about them, who sometimes refused uniform in favour of oilskins. By May 1941, one vessel of the RNPS was lost every two days, usually because of bombing.[18] They were armed although relatively lightly. A 12-pounder gun on the foredeck, a couple of Lewis guns, an Oerlikon – heavier than 12-pounders – too. Some machine guns. The vessels hunting submarines had ASDIC, a kind of sonar. The minesweepers also had rifles because they needed to shoot at mines.

If a minesweeping trawler did its job, the mines would float to the surface, hopefully far enough away from the trawler. Sometimes they didn't and a trawler went up or it went down. If it survived the floating mine, the men fired armour-piercing bullets. 'One or two exploded,' said Joe Steele, a docker from Liverpool who served as a signaller for the whole war, 'but mostly we'd sink them.' The interviewer asks him what he would aim at. 'Not the horns. Just put holes in it and sink it.'[19] Shooting the contact horns – these mines really did look like the cartoon versions – risked the mine and ship exploding together.

In 1939, Prime Minister Winston Churchill wrote to the Fourth Sea Lord, the naval official in charge of supplies. 'I am told that the RNPS/Minesweepers men have no badge. If this is so it must be remedied at once.'[20] The men of the Merchant Navy – a fighting force made from commercial cargo vessels – were not given a badge for decades, and with no uniform either were often, when ashore, given white feathers (to signify they were cowards) by unkind civilians. RNPS men who had served six months at sea now got a handsome silver badge featuring a shark pierced by a marlinspike, what looks like a fishing net (it is actually a boom defence net, usually strung across a harbour entrance), and two mines. The shark was meant to signify a submarine; the mines need no explanation.

Charles George Stoakes had served in the First World War and, because he was a Royal Navy reservist, immediately reported for duty for the Second. He was assigned to the *Lady Elsa*, previously a yacht and now a fighting minor vessel carrying 26 men. 'At first it was the simple moored mine the enemy was laying, but before long the Germans had perfected the sinister magnetic mine, which followed the acoustic mine. The real headache came when the enemy started laying a cocktail of all three mines. To fight them, all these "minor war vessels" were festooned with wires, sound apparatus, and electric cables. This was the nightmare these men had to face each night, and every day, which required nerves of steel.' On 19 November 1942, Stoakes' vessel *Ullswater*, a former whaler, was blown up by a torpedo off Portsmouth. There were no survivors.[21]

The *Lady Shirley*, previously a Hull trawler, was one of the Harry Tate superstars because she captured a U-boat. Lieutenant-Commander A. H. Callaway described events: A glimpse of a U-boat conning tower triggered the *Lady Shirley* to drop depth charges. A periscope appeared, and then Germans standing on the conning tower. *Lady Shirley* fired at them. Two officers were immediately killed, and soon the U-boat crew, 'finding themselves getting knocked about' by shells, surrendered. Because the *Lady Shirley* was a fishing vessel and not a Nazi cruiser, she collected survivors, and the crew stood to attention for a burial ceremony for the dead Germans. 'Asked if any prisoners showed aggressiveness, Commander Callaway said, "only one or two, but all were dumbfounded and bamboozled that such a tiny ship could have destroyed their U-boat."'[22] Callaway was awarded a Distinguished Service Order. Two months later, the *Lady Shirley* was patrolling near Gibraltar when U-374 torpedoed her. All 33 crew went down with the ship.

In the lists of trawler and drifter casualties, the progress of the war is evident. In 1940, British troops were trying to flee Namsos in northern Norway. This was known as the first Dunkirk. A fleet of fishing vessels named after football teams arrived to do antisubmarine patrols. *Aston Villa*, a Grimsby trawler, was joined by *Blackburn Rovers*, *Leicester City*, and *Huddersfield Town*. The trawlers were stuck in the fjord for ten days under aerial bombardment. They learned that staying in the shade of the fjord could hide them from planes and they could attack the planes before they were spotted. East side in the morning, west side in the evening. Stay west until the 03:00 reconnaissance plane appeared, which meant an hour until the bombers arrived, then head back to the east.

They also decided to use the local fir trees. 'Wherever possible,' write Lund and Ludlam, 'crews went briefly ashore to chop down small firs . . . all the greenery was then draped over the decks and rigging as a camouflage against the searching bombers. Soon a number of trawlers were steaming about the fjords looking like mobile Christmas trees.' A passing sloop that signalled 'the decorations look pretty, when is the wedding' got short shrift.

On 20 April, the Germans bombed Namsos. The town was mostly wooden and it was obliterated. Its people had fled and casualties were light, but the inferno was appalling. Even so, HMT *Arab* headed for the quay and attempted to put out the fires. In the fjord, HMT *Rutlandshire*, a Grimsby fishing boat, was the only vessel in sight when the planes arrived. Twenty of them in this narrow fjord. 'We gave them all we knew,' chief engineer Joseph Winner told the *Daily Sketch*. 'At one time there were 16 machines swooping on us from different angles.' These fighting fishermen – and naval personnel and everyone else – lost more vessels than any other branch of the Navy and showed as much courage. In Lowestoft, 2,400 are remembered on a memorial. The dead of Harry Tate's Navy, who have 'no grave but the sea'.[23]

A Ministry of Food advertisement in 1940:

>The fishermen are saving lives
>By sweeping seas for mines
>So you'll not grumble, 'what, no fish?'
>When you have read these lines.[24]

A fish processor at Mballing, Senegal

9

Before, There Was Fish

God said to them, 'Be fruitful, multiply, fill the earth, and subdue it. Have dominion over the fish of the sea, over the birds of the sky, and over every living thing that moves on the earth.'

GENESIS 1:28

In tourist guidebooks to Senegal, intrepid visitors are advised to head to the fish market in Mbour. This is a kilometre up the beach from my hotel, where I have been sitting on the peaceful terrace high above the sand watching fishing boats in the sea. With borrowed binoculars I watch the pirogues – heavy wooden canoes – casting nets then circling and circling, and I wonder what they are catching so close to shore. These are small pirogues that seat fewer than half a dozen fishermen. Bigger ones are further along the beach in the fishing village, the place I have been advised to avoid and especially at night. I watch a solo fisherman bashing a cosh onto his boat again and again and again. It looks like peculiar violence, like he is furious at the sea. Later I understand he is scaring fish up into his net.

I have come to Senegal to attend a workshop hosted by CAOPA, a pan-African organization defending artisanal fishing, but today is Sunday and I have no meetings planned. The weather is heating up fast and will be over 40 degrees soon, and I don't want to lie around wasting a day, seduced by air conditioning. I'll go to the fish market, although I am alone and nervous about it: Mbour is not the tourist zone of Saly, where the workshop took place and where there were plenty of tourists. But I have to do something useful, and I refuse to be anxious about going to a market because so far in Senegal I have encountered nothing but kindness. I climb down the steep steps to the beach, arriving next to a place I call Sporty Sandy Hill, where young men even in this early heat are pelting up and down or hopping and jumping like football players warming up. It is impressive. This is the incubation zone for the next Mané, Senegal's superstar football player, just as every football pitch and scrap of waste ground is growing the next best export. Better to be a Mané than a fisherman. Better to be in Europe than in Senegal.

A man approaches me quickly, descending too, probably from a hotel. His T-shirt reads '*oui je parle français*' and he introduces himself. Marcel, tour guide. He tries to interest me in things that I have already read about: Jaoul, where the first president of the republic is from (yes I know); the attractions of Bandia animal reserve (yes I know); that slavery actually started in Senegal. Here I stop him and say, too abruptly, 'I'm not a tourist. I want to know about fishing.' 'Sure!' he says. 'My family are fishermen. Everyone I know is a fisherman. For 10 euros, I can take you to the fish market.' This seems like a bargain.

The first signs we are nearing the market are the lines of pirogues parked on the beach. They are big, heavy, wooden. Their colours are garish, the designs are rough, the kind of rough that art gallerists would call 'naive'. Some pirogues have green, red, and yellow Senegalese flags, small like bunting. On the side, the owners paint their name, the year of the pirogue, and often a marabout or imam, Muslim holy men. On one, there is a mermaid with peculiar breasts. Marcel now tells me a long and sad story about a friend of his who took a pirogue that was sailing to the Canary Islands. For years, so many pirogues took so many illegal immigrants to the Canary Islands, and so many Senegalese drowned, the route became known as *Barça o Barzakh*. Barcelona or death, in Senegal's majority language Wolof. Marcel tells me the story with what seems to be genuine grief, because his friend drowned and the friend's mother now weeps when she sees Marcel because she is reminded of her lost son. When I apologize and say my recorder wasn't recording, he repeats it perfectly. Same intonation, same words, same grief.

Before, There Was Fish

Just beyond the pirogue park we stop because of the smell. There used to be fish smoking done here so that the fresh fish could be preserved despite no refrigeration, and the huts are still standing, almost. Most of the fish processing and smoking has been moved a few kilometres away to a village called Mballing because neighbours here hated the stench. Even so, a few women are still here, cutting and gutting. Here is Senabou Sene, gorgeous in a bright-red striped robe, covered with cloth from head to foot, and sitting under an umbrella that could never be strong enough for this sun. She has three children and has been 'transforming' for three years. The French for 'fish processor' is the lyrical *transformatrice*, and it is always feminine because nearly all fish processors in Senegal are female.

Senabou's husband is a fisherman and right now he is at sea with her eldest son. She buys fish from the pirogues that land 200 metres away, then guts and slices them; her colleague will put the fish through a mincer and sell it in sachets for 500 CFA or less than $1 (CFA is the franc of West African Communities). It's a living, she says, but these days sometimes there are no fish. Luckily there is a generous credit system at work. She can buy fish and pay the fishermen back when the fishing is good. Or she takes from the family savings and pays herself back. It is not easy and she would like to leave. Senabou knows seven women who departed on the pirogues that still – though more secretly than before – take migrants to the Canary Islands. She didn't have the 4,000 CFA ($6) for the ticket. 'I saw them leave and I wanted to go with them.'

Marcel says he knew a woman who left when she was pregnant and she gave birth 'over there'. Any time I mention the pirogue emigration, Marcel is quick to say, 'It's forbidden.' He says that now you get fined 1,000 euros ($1,136) and get three years in prison. The big pirogues can take 100 people. It takes six days to get to the Canaries. Some of the pirogues we walk past were seized, and Customs has sold them to new owners, who have repainted them. Senegal is routinely in the top five sub-Saharan countries spitting out their people northward in hope. Senegal's unemployment rate is just above 3 per cent but in young people it is 16 per cent,[1] and two-thirds of Senegalese are under 25. All these numbers can add up to a big push: outward, anywhere except here.

Illegal migration is forbidden but it persists. Although any number must be imprecise, there are organizations that still attempt to calculate how many Senegalese die on the Canaries route. Usually the numbers are provided by survivors: for example, the 50 dead in a pirogue that arrived in Mauritania in late 2020. The pirogue had broken down, and 'one by one', a Canaries news site reported, 'the passengers died while trying to get closer to land to seek help'. By October 2020, the International Organization for Migration calculated that 11,000 people had arrived in the Canaries, compared to 226 the year before.[2] This was nothing like 2018, when more than 60,000 arrived,[3] still substantial numbers for something that the government has outlawed.

In August 2023, 15 drowned corpses were returned to a beach near Dakar's greatest mosque. They had set off in a pirogue, and they were young. 'That's what hits me hardest,' a local fisherman who had gathered a corpse told French TV. 'The youngster. He was 17, 18.' In early July, a pirogue with 100 people departed from Fass Boye and sank. Only 38 survived. Two of the 60 dead were brothers-in-law of Amedi Dieye. 'Young people spend months at sea, only to return empty-handed,' he said. 'The authorities have sold off all our resources so they are responsible for this tragedy.'[4]

A young man in the same village was undeterred. The president has sold the sea to foreign boats 'that tire us out,' he said. 'If I can find a pirogue this very evening, I'll get on board.' He does not mean a pirogue to go fishing. There's no point to that anymore. If there were any fishing, they would not need to escape.[5]

The closer we come to the fish market, the darker the air gets with smell and dust and fish guts. Before we reach the worst of it, I turn my face to the sea to watch a pirogue landing its catch. The wooden boat is huge and about two dozen fishermen are pulling the net onto the shore, heaving in a balletic line. I see no lifejackets, but Marcel says they will be in the boat because they are precious. I tell him I've seen none being worn by the fishermen I have been watching from the hotel and he says, oh, that's different, it's near the shore, it's only 10 metres deep there. Marcel has a strangely precise number of how many fishermen are drowned at sea each year: 6,200. Why not? It is believable and unprovable.

The fishermen have been yelling at us, but Marcel won't translate. A disembarked fisherman approaches. His black baseball cap reads Palm Springs, and other than that he is wearing only shorts. Around his neck are silver chains and his eyes are red. He asks if he can marry me, if he can take me fishing. He seems to think that is the same thing. I know this is nonsense because no Senegalese fisherman would take a woman to sea. We are bad luck. Marcel hurries me away. 'He is drugged,' he says. 'They have to be drugged or drunk to stand the life at sea. You cannot trust them.'

He is not dismissive of all fishermen: he is one too. He says everyone in Mbour is involved in fishing. His father is a fisherman, also his brothers. He goes sometimes to fish for *lotte* (monkfish) or barracuda. This week he will probably go on Wednesday. He knows his fishes, easily identifying the ones on sale on the open market stalls, where they are besieged by flies. Speedy porters sprint past carrying plastic crates of fishes on their heads. Now and then a fish drops from the crate and it is always quickly retrieved by a stallholder or a dog. It will never go to waste. A small boy says something loudly at me in passing and I don't understand him and assume it is hostile. Actually he is telling me that my phone is hanging out of my pocket and to take care, and I am ashamed.

In the market, fish stalls are operated by both men and women. I ask about the fish on display. There is *thiof* or *lotte* or barracuda or *dorade*. *Thiof* (pronounced 'choff') is a white grouper, *Epinephelus aeneus*, and according to the IUCN Red List, its population is 'near threatened'. What Senegalese know is that they can't get hold of it as they used to be able to, and when they can it is too dear. *Thiof* is supposed to form the basis of Senegal's national dish, *thieboudienne*. The dish's name translates from Wolof as 'rice of fish', and it is made with spices, fish, rice, tomato sauce, onions, carrots, greens, cassava, and peanut oil. It is usually served from a large communal bowl and you help yourself with your fingers. You can't be aloof when you are eating *thieboudienne*.

Off Senegal, *thiof* have gone. Either they have moved or their population has collapsed. Fish biologists call the desperate situation 'a loss of biomass'. I prefer the sadness of Dakar nutritionist Codou Kébé, who said that 'the sea no longer supports the weight that is loaded on it'.[6] *Thiof* was called 'false cod' by European fishermen,[7] probably because it was tasty and abundant. The *thiof*'s biology is nothing like cod's, and that has contributed to its decline. Born female, it turns into a male at about 12 years old or when it reaches 80 cm in length. Because fishermen seek to catch the biggest fish, they have fished all the males and the females cannot reproduce.

Here at the fish market, there seems to be no lack of fish. I look out to sea and know that the important fish are missing, though here fishes lie on grubby stalls or on the filthy ground in good enough numbers. There are *lotte* and *rougets* and unidentified yellow ones. There is a small heap of rays, an animal that is among the most threatened fish: 41 per cent of 611 ray species are at risk or critically at risk.[8] Round the back of the stalls I see sharks under plastic; 36 per cent of 536 shark species are in a critical state.[9]

Behind the market hall, refrigerated trucks wait for cargo. They will then head inland to this country and others, or to fish-processing factories that will process and freeze for European markets. The only sign of refrigeration on this market, intended for locals, is a fridge freezer connected to nothing and used only as a box. This is partly why sardinella and another small pelagic, *bonga*, became staple foods: they can be processed and salted and dried and they will keep. The Senegalese are fish people: fish make up nearly half of the animal protein they consume,[10] and until sardinella began to be exported, small pelagic fish constituted 80 per cent of fish landings.[11] Because of its abundance, sardinella became known as 'the people's fish', and it was the cheapest source of protein for the poor. By 2018, the consumption of small pelagics had halved to 9 kg, and by now it will be smaller.[12] The sea cannot bear the weight of our appetites.

The pirogues must go further and further out now to find fish. Everyone I talk to blames the big boats. Everyone blames the foreigners. This is a common lament anywhere in the world, but here it is mostly accurate. Industrial trawlers have been fishing in Senegal's rich waters for years, arriving from different countries at different periods to fish different fish. Pirogue fishermen talk of trawlers given permits by the Senegalese government – the big boats that sit out at sea like a waiting army, taking the Senegalese small pelagics. And not just pelagics. Under the EU's 2019 fishing agreement with Senegal, which expired in November 2024, France, Spain, and Portugal can hunt 10,000 tonnes of tuna a year using 28 freezer tuna purse seiners, 10 pole-and-line vessels, and five long liners.[13] Two Spanish trawlers are also permitted to fish for 1,750 tonnes of black hake per year.[14] For these permits, the EU pays €1.7 million ($1.8 million) per year.[15] This is a bargain. According to a paper by the fisheries researcher Dyhia Belhabib and colleagues, the EU was paying under a tenth of the value of the fish landed by its vessels. China has an even better deal. Under its bilateral agreements with Mauritania, Senegal, and other West African countries, it only pays 4 per cent of the value of its catch.[16]

Senegal also has a fleet of trawlers that target demersal species such as hake and *thiof*, fish that live and feed near the seabed. This fleet has done its job so efficiently that now there are hardly any hake and *thiof*. Nobody – the EU, the Chinese, the Turkish, the Senegalese industrial trawlers – is meant to be catching small pelagics. Yet every Senegalese fisherman and fish processor and fish merchant and fish porter will tell you that they are.

Abdoulaye Ndiaye is the secretary general of UNAPAS, an organization representing small-scale fishermen. Abdoulaye is a tall and tired man. He has elegance and fire in his speech even though, as is the case with everyone I speak to in Senegal, if he talks to me about fishing, he must talk of loss and dismay. His words are big and all the more alarming for the measured way he uses them. *Penurie. Famine. Carence.* The words – shortage, famine, deficiency – sound more epic in French and therefore they are more accurate.

He also talks carefully of governance and how it is failing. There is illegal fishing off Senegal, but the problem is also the legal fishing. Everyone is seeking the same dwindling fish, and meanwhile sardinella and bongo, Senegal's staple foods, have been moving north at the rate of 112 miles per decade since 1995, probably because of what scientists call 'strong warming', in the seas off Africa's north-western coastline.[17]

Not knowing what is happening is usual when you are trying to understand who is fishing what off Senegal. Until 2024, the Senegal Fisheries Ministry refused to publish lists of foreign vessels it had granted licences to. In 2020, local and international organizations protested when the Senegalese government planned to grant 52 licences, to mostly Chinese-owned vessels, to fish for small pelagics as well as everything else.[18] This was the year after the FAO recommended that the small-pelagics fishery be reduced by 50 per cent to give it even a small chance of recovery.[19] Even though Senegal was in the midst of Covid, the applications were noticed and objected to, enough that the government publicly backed down, probably because Greenpeace published a report that showed some of the vessels had been implicated in illegal fishing. Later, the government issued some of the permits anyway, to the murkiest of the boats. When the licences were finally made public, half of the vessels had Chinese names but their owners were listed as Senegalese. Among the 19 foreign vessels listed, none had Chinese or Taiwanese owners, at least on paper.

Ndiaye does not think things will get better. He offers absurdities: the government continues to license large vessels but the country's only scientific research vessel broke down in 2015, so they have no official idea of the health of fish stocks. The government spent millions moving the fish processors to sites such as Mballing, but they spent all that money and the women have no fish to process.

Before, There Was Fish

Without a research vessel or at-sea monitoring, there are only eye-witnesses and anecdotal reports. Pirogues encounter trawlers either because the pirogues now have to fish further away or because the trawlers illegally come at night into Senegalese waters. The pirogue fishermen note the trawler names and numbers and report them, and nothing happens, says Ndiaye. There are battles at sea between these big boats and the pirogues. Senabou the *transformatrice* had heard stories of pirogues fighting trawlers. I asked her how they fight when there are two boats and water between them: bows and arrows? Marcel stepped in with his explanation. 'I have seen the boats at night,' he says, 'when I am fishing. Their huge lights.' The fighting is done by ramming sometimes. The trawlers, huge next to the pirogues, switch on their engines and ram them. Sometimes the pirogue crew throw plastic drums at the trawlers. It's not a cannonball but it is all they have. What they cannot do is throw plastic drums at the places that many Senegalese also blame for their troubles, their emptying seas, their missing fish. They cannot hit the fishmeal factories.

It is an early start, the world still dark, because there is an hour's trip ahead and no one is sure of the roads. Not the roads themselves, because the highway between Dakar and the important city of Thiès was built by the same companies who built the French autoroutes, so that whenever I am on a motorway in Senegal I get briefly confused and think I'm in Brittany. Same signage, same quality of construction, same everything. We can be sure of the quality of the main roads, but people are unsure of everything else.

We are four in the car: two staffers from Greenpeace's Dakar office, Amagor and Kaly, the driver, and me. After initial tired morning talk, they tell me that trouble is expected. A famous opposition politician has been accused of rape and his hearing is scheduled for after lunch. It is the second hearing in his case, and after the first one there had been riots and people had been killed. It is so early that there is no chance we'll run into trouble on the way out of Dakar, but they can't assure me of what may happen on the way back. Look, Amagor says, pointing out of the window at an underpass. There are military trucks and armed men standing bored behind plastic riot shields. They are getting ready.

We are heading for a different court, and proceedings will start at 8 a.m. Today Greenpeace is expecting a judgment in a case that has been in the system for many months. The year before, the Taxawu Kayar Collective, a cooperative of fishermen and fish processors from the fishing port of Kayar, filed a case against a factory near their town, up the coast from Dakar. The factory used to belong to a Spanish company called Barna and had been recently sold to the Senegalese-owned Touba Protéine Marine (TPM). Touba is a word with great meaning in Senegal: Touba is the holy city of Mouridism, a Sufi brotherhood. For a company to call itself Touba was to say something about its affiliations and connections.

He's a showman, Amagor says, with some contempt. He means the owner of TPM, a man named Babacar Diallo. Amagor has tales about the showman's love of show. How he dresses up in religious clothing, turbans, and flowing robes to show his affinity with Mouridism and Touba. How he once told a TV reporter that he slept in his office in the factory Monday to Friday and how could he do that if it stank like people said it did? How once when people were protesting outside his factory he bused in a load of children and villagers from nearby villages and gave them T-shirts and made the news with his happy factory folk.

Why would there be protests outside a nondescript low factory building on a road leading out of a small fishing port? Because TPM is a fishmeal factory, and the kindest way I can describe the fishmeal industry is to say that it is intriguing.

The people of the Taxawu Kayar Collective are certain that TPM has stolen their fish, ruined their livelihoods, and polluted their local lake by dumping toxic wastewater into it.[20] 'I get my water in the next town now,' a taxi driver told a reporter from *Hakai Magazine*.[21] The collective has footage of a truck tipping waste into Lake Mbane, and a laboratory at the University of Dakar found illegal levels of chromium and selenium in the lake water, and also in Kayar's tap water. The factory, said

Amadou Kamara of Kayar, 'affects our health because of its olfactory nuisance. All the people suffering from respiratory diseases live with difficulty in the village. The health of pregnant women is also affected.'

'I have a bruised heart,' Kamara continued. 'My right to a healthy environment scorned, and trampled, my health and my economy threatened.'

The collective is organized enough to attract attention and to have successfully proven that its cause is good, so Greenpeace has supported it. This is an unprecedented case in a country where hardly any environmental prosecutions are even brought. No one has sued the fishmeal industry before.

Take a fish. Eat its flesh if you want, but don't scorn the rest. It can always be used, one way or another, whether you want fishmeal or face cream.

Cod skin can become a graft for a badly burned man such as Pétur Oddsson, an Icelandic power station worker who received a 60,000-volt shock in 2020 that burned half of his body. His skin is now scaled, because as Oddsson lacked enough skin on his own body to graft, ingenious people at the Icelandic company Kerecis took the hardy skin of a cod and used 7,000 square centimetres of that instead. After the Nagorno-Karabakh war in Azerbaijan in 2020, two surgeons – one Icelandic and the other British – travelled to the war zone and used cod skin on patients who had been burned with white phosphorus dropped by drones. 'This new style of drone warfare,' wrote the surgeons, 'has resulted in more burn and flash burn injuries than those previously seen in traditional combat.' They used cod skin produced by Kerecis. The Kerecis website has some remarkable stories: a 20-year-old wound that had never closed but healed in two months after a fish skin graft. The product 'recruits the body's own cells, which are then incorporated into the damaged area and ultimately are converted into functional, living tissue'.[22]

It doesn't have to be cod. In Brazil, where fish skin grafts were pioneered, Maria Ines Candido da Silva was burned by an exploding gas canister in her workplace. She healed with tilapia skin.[23] In 2017, after terrible wildfires, veterinarians at the University of California, Davis, used tilapia skin to treat two burned bears and a mountain lion. Tilapia skins 'supply collagen, hasten healing and are harmless to the animals if swallowed'. The bears and lion were fixed and tilapia has since been used to heal kittens and ponies. A photo of a kitten that survived a fire shows his paws thickly wrapped in bandages. If that kitten is anything like my cat, the fish skin would soon have been dinner. When Olivia, a Boston terrier, was treated with fish skin, the effects were striking. She had been 'pretty mopey' before the tilapia skin, said her human guardian Curtis Stark. 'It was a day and night difference. She got up on the bed and did a back flip.'[24]

Kerecis's product is a 'biologic, acellular matrix derived from the skin of wild-caught Atlantic cod that contains intact epidermis and dermis layers'. It has a three-year shelf life and can be revitalized by being dipped in a saline solution. The fatty acids it contains have good anti-inflammatory and bacteria-limiting properties, and the skin is easy to use, being robust and stretchable. Even better, rather than just 'buying time' by covering horrible burns, the fish skin triggers the healing process.[25]

Products derived from fish turn up everywhere. Guanine, an ingredient produced from fish scales, is found in bath foam, perfume, nail products, shampoo, skincare products, and lipstick. Fishes can become gelatine, glue, and squalene. They can be in your dietary supplements or your glass of wine: isinglass, derived from dried fish air bladders, is used in wine-making.

Futuristic piscigrafts are irrelevant to most people in Senegal, but they still have to deal with the fish by-products industry, because they have to watch as fish they could eat are transformed into the most common and lucrative fish by-products and sent abroad.

Fishmeal is a sandy grubby powder that is worth billions. Minks eat it. Pigs eat it. Cattle eat it. Fishes eat it. It is also an ingredient in fertilizer and pet food. Fishmeal is part of a huge silent industry of fish by-products that used to be known as FMFO (fishmeal and fish oil) and has now been rebranded by the industry association as 'marine ingredients', a vague term that covers a broad range of fish-derived products, most of which are never noticed.

The Food and Agriculture Organization defines fishmeal as 'the clean, ground, dried tissue of undecomposed whole fish or fish cuttings, either or both, without the extraction of part of the oil'.[26] American pets may also find themselves eating fish liver and glandular meal, made from dried complete fish viscera; more general fish by-products, which are 'non-rendered, clean, undecomposed portions of fish (such as, although not limited to, heads, fins, tails, ends, skin, bone and viscera)'; or condensed fish protein digest, the 'condensed enzymatic digest of clean undecomposed whole fish or fish cuttings using the enzyme hydrolysis process'. 'Undecomposed' is doing a lot of work here, a strange term because unless a fish is caught by a vessel with freezer capacity – which no artisanal fisher has – it will be decomposing for days.[27]

In theory, fishmeal is impeccably sustainable. It should be a product made from parts of fishes that would otherwise be discarded. It is because of fishmeal that the term 'reduction fisheries' was coined. Take the meat and the prized flesh, and the rest can be reduced or diverted to other uses. That is the theory. It is not what happens in West Africa.

Each year, half a million tonnes of wild fish are caught in African waters and sent to Europe to feed animals, when they could have fed 33 million Africans instead.[28] That number of people is more than the combined populations of Senegal, Mauritania, and The Gambia. Globally a fifth of wild fishes become fishmeal or fish oil,[29] but 90 per cent of fishmeal fishes are suitable for human consumption.[30] China uses nearly half of all FMFO, but then China uses most of everything.[31] Europe also takes and takes. Spain imports the most fish oil from Senegal, and France takes 70 per cent of Mauritania's.

In 2019, Changing Markets Foundation produced a detailed investigative report into the fishmeal industry in India, Vietnam, and The

Gambia. In each country, the pattern was similar: first the foreign factories come, then they take the fish, then they pollute the environment, then the fishermen say there are no fish left. Although noting 'a surprising lack of traceability and transparency regarding the origin and supply chain of farmed seafood products', the foundation linked plenty of reputable European retailers to FMFO supply chains that may use West African fish. For example: Sainsbury's, ALDI, Lidl, and Marks & Spencer in the UK; Auchan, Groupe Casino, Monoprix, and Système U in France (that is a full house of big French supermarkets); all the German ALDIs; and Mercadona in Spain.[32] Sainsbury's, Marks & Spencer, and Mercadona have said they don't use West African FMFO in their seafood products, but how will they know? The European Union's IUU regulations, which require a catch certificate or similar to show the fish was legally caught, do not cover fishmeal.[33]

In Mauritania, some Chinese-owned fishmeal factories have their own trawler fleets. The fish bypasses human mouths and goes straight to the factory. That is one tactic. The other is to outbid. The fish processors tell me that they cannot compete with the prices that fish factories will pay for sardinella. Fish processors get no subsidies from the government, unlike Chinese fishing companies. In The Gambia, one fishmeal factory used 40 per cent of the country's entire fish catch in one year.[34]

Black and white. White stealing from Black. The villain/victim template seems neat and unassailable. It is true – Europeans are using fish that West Africans could eat – and yet it is not all the truth. In Senegal, The Gambia, Mauritania, local fishermen supply fishmeal factories because the factories pay better. In Senegal, a fisheries union official tells me that there are too many pirogues. They are chasing smaller and smaller fish to sell to the factories. It is unsustainable.

The website of the IFFO, which used to stand for International Fishmeal and Fish Oil Organization and now somehow is short for the Marine Ingredients Organisation, is plain. Marine ingredients 'play an essential role in global food security'.[35] Rubbish, said Francisco Mari of Bread for the World at the CAOPA workshop. 'It's just for prawns and salmon – they are not necessary foods; it's a luxury. It's not enough for us to have 10–15 per cent of fish to feed salmon and prawns. Indirectly, we are eating your fish. It's not essential for the alimentation of the US or Japan. We are taking your fish to feed a luxury product that we don't need.'

In Thiès the courtroom slowly fills up, and I decide that the women in the row in front of me, dressed magnificently, are fish processors and fishwives. A nice man in uniform passes down the aisle and says very politely, please don't use your phones, but I think he is less nice when he approaches me and asks me directly to do the same, although mine is in my bag and switched off and everyone else has theirs in plain sight. Once when I was reporting in the Old Bailey, a phone rang and everyone looked at me, the only person on the press benches. The prosecuting lawyer was the guilty one.

Finally a man in judge's robes arrives from a side door, sits down, and mumbles. The microphone isn't working. No one can hear him, and no one goes up to fix the microphone. I look around in puzzlement but there is no unrest and no disquiet. After 10 minutes, a court official goes to the judge and signals that he should use the other microphone. After that I watch the procedures and understand nothing. Advocates advocate and argue. Witnesses are called. I hear nothing that makes me think this has anything to do with fishing.

Outside, I see a group of journalists talking to a lawyer. I think it's the factory's lawyer so I go over, but they are speaking Wolof and I have no translator. Back to the courtroom for more bafflement, past a ridiculously huge and ostentatious Infiniti SUV, which someone tells me belongs to the judge. Then Amagor comes and whispers with urgency: 'We are leaving. Now.'

I'm bewildered.

'Why?'

'He already did the judgment. We lost.'

'When? How did I miss it?'

It was the inaudible droning right at the start. And now I think the judge did it deliberately, that he knew that the microphone was off, and that he thought that delivering his judgment in favour of the factory would cause less ruckus if no one could hear it. The lawyer speaking in Wolof did not have supersonic hearing: he had been told of the judgment in advance.

The Greenpeace team are in a hurry now. There is no point staying, and it is safer to get to Dakar before the opposition politician's court case finishes and trouble starts. All the Kayar people have gone. They have better things to do, like make a living.

In the car, some things become clearer. The judge knew his judgment was inaudible and refused to repeat it because not all the lawyers were present. The missing lawyers were Greenpeace's, and because Greenpeace had no lawyer present, they couldn't immediately get a copy of the judgment either. This seems like an inexplicable error after two years of effective and assiduous campaigning. It seems a shame.

In 2010, the Mbour authorities moved the fish-processing women away from the fish market to Mballing, a short distance up the coast. There was a fishmeal factory here that exploded a few years ago. Other factories freeze fish. The Mballing I see is shacks and dust and piles of carefully stacked fishes. The women were moved here because there was a plan to build a proper harbour so that the pirogues did not have to be lugged down the beach using only logs and strength. The plan is still a plan. The town still has no harbour.

Anta Diouf, descendant of generations of transformatrices, leads the fish processors. Her energy is formidable but also sad. 'Where we used to have 50 boxes of fish to process, today we can't even get 20 boxes. This is the terrible consequence of the trawlers overfishing, and of the fishmeal factories selling our fish so they can be used as animal feed thousands of miles away. When I remember those beautiful moments at the fishing dock with endless fish, and then I think about the current situation, I have tears in my eyes.'

Before, There Was Fish

Anta leads me to meet a group of women resting in the shade. They span all generations, and they are resting because there is nothing for them to do. The fish are smaller and scarcer and they cost more. Sometimes the women have to wait days for any fish to work with. Mesa Diaba Faye is 83. I ask her how things have changed since she started working. It is very hot, and the women all look as lethargic as I feel, but her answer still cuts like glass.

'Before, there was fish.'

© Open Seas

Sea Beaver

10

Down Below

The best way to observe a fish is to become a fish.

JACQUES COUSTEAU

How do you police the ocean? There are no squad cars on the high seas; most crimes happen far from shore in international waters, where no one is responsible for oversight. It is hard enough monitoring what goes on above the water. When wrong things happen underwater, how do you detect them?

You swim like a fish.

Science can be like war, with long periods of boring waiting, short bursts of activity, and infuriating tech. I have this unoriginal thought during my second day on *Sea Beaver*, as we are berthed at Tobermory on the Isle of Mull, an inner Hebridean island off western Scotland with colourful harbour-front houses and a famous ginger town cat. This trip was a long time in the planning. For nearly a year I had been talking to Greenpeace about going out on its vessel *Sea Beaver*, a converted supply ship that they were using to monitor marine protected areas. One or two possible trips came and went. Engine trouble. More engine trouble. Finally Greenpeace said, 'We are lending our boat to a Scottish NGO: talk to them instead.' They passed me on like a fish in a net swung from boat to boat.

The Scottish NGO Open Seas was founded in 2015 and began, says head of communications and campaigns Nick Underdown, 'as two part-timers'. Its aim is to protect coastal waters and communities and to promote sustainable seafood through research and campaigns. It does this partly with a smart and thoroughly researched website. But a website is not a boat. Open Seas can't afford one, so it borrowed a research vessel from Greenpeace and launched a six-month survey of the seabed around Scotland's inshore waters with remotely operated underwater vehicles (ROVs). The mission: to prove that the gear used for bottom-contact fishing – particularly scallop dredging – was damaging the seabed. It is extraordinary that this evidence was still lacking in 2022. Underwater research is expensive and frustrating and imperfect. To monitor the inaccessible, you can only use humans or remote technology. Humans are expensive and diving is never completely safe. ROVs are a boon. They provide an exceptional opportunity to add to a patchy body of data about what humans have done to the seabed. If they work.

I meet the *Sea Beaver* at Oban. Before I have been properly introduced to the crew, marine biologist Rohan Holt starts showing me footage on his MacBook Pro. 'Look,' he says. He shows me crisscrossing tracks on the seabed. That's what dredgers leave behind. He sounds sickened because the tracks are shocking, and even worse is the red rain: small particles of dark red things are floating everywhere. It is entrancing and sinister because it is the annihilated remains of maerl, a seaweed as hard as coral that grows barely at all per year. 'It's a Cnidarian,' Rohan tells me, a species that sounds science fictional. Maerl beds are excellent habitats for fish and invertebrates. Their combination of polyps and hard branches makes for an attractive environment with plenty of tiny spaces to hide in, which are flushed by oxygenated water. Nature Scot, Scotland's government nature agency, calls maerl beds 'underwater carpets'.[1]

Maerl beds are supposed to be protected. They are one of 81 'priority marine features' designated as such by the Scottish government in 2014, along with minke whales, flame shells, fireworks anemones, and native oysters.[2] Their protection consists of being inside Scotland's 247 MPAs.[3]

Marine protected areas are a method of ocean conservation that aims to limit human activity. The UN defines an MPA as 'a geographically defined marine area that is designated and managed to achieve specific long-term biological diversity conservation objectives and may allow, where appropriate, sustainable use provided it is consistent with the conservation objectives'. This label conceals a massive amount of variability. Any government or authority can establish an MPA, and 'protected' can mean whatever they want it to mean. A good one, in the words of marine conservationist Callum Roberts, is 'leaky': its positive effects spread outside its boundaries. Protect the fishes inside it, and they live longer and grow bigger and produce vastly more eggs, meaning more fishes everywhere.

In 2018, Boris Worm examined fishing activity in the 29 per cent of European seas that are protected in some way. Trawling, the most damaging form of industrial fishing, was 40 per cent higher inside the MPAs than outside them.[4] Then Oceana studied 73 of Britain's MPAs and found that bottom trawling and dredging took place in 71, legally. More than 44,000 hours of fishing using bottom-contact gear had happened in all but two of Scotland's MPAs.[5] South Arran MPA – whose category is 'biological' – is meant to protect ocean quahog, kelp and seagrass, maerl beds, and burrowing bivalves. In 2020, fishing vessels dredged and trawled there for 2,295 hours, an underestimate because that guess is based on incomplete AIS data.[6] In 2023, Veronica Relano and Daniel Pauly established a 'paper park index', using various criteria to decide whether an MPA was protected only on paper. Of the 184 MPAs they studied, they concluded that a third qualified as paper parks.[7]

Mark Costello, who is running an EU project to assess where best to put MPAs for maximum effect, calculates that only a quarter of coastal countries have any marine reserves and only 1 per cent of the ocean is without fishing. We don't have enough protected areas and the ones we have don't work.

*

In Oban, the current mission is to wait. Open Seas' ROV broke, a new one has been ordered, and the crew are waiting for delivery and then will figure out how to assemble it. This, they say, will take three hours and then we will go to sea. It does not take three hours and we never really go to sea.

The crew are Kim, the pilot; Rohan; Theo, a recent marine biologist graduate whose middle name is Apollo and whose confidence is as vivid as his tattoos; and Neil, a diver employed as the boat engineer.

The new ROV arrives. It is unpacked and peered at, and the assembly and understanding begin. Even without a functioning ROV, we depart from Oban and head to the Isle of Mull. I note a beautiful coastline, and someone says, 'That's probably Hugh's. He founded Open Seas.' Good for Hugh. We depart with hope that the technology will not fail us. Instead, I go from waiting on the boat at Oban pier to waiting on the boat at Tobermory pier.

To pass even more time, Rohan shows me footage of a scallop swimming. It is astonishing. It's not really swimming but propelling itself by opening and closing its shell. Rohan says, 'It's like a pair of false teeth swimming around.' The Japanese name for scallop translates as 'full-sail fish'.[8] I gather scallop facts. They have hundreds of eyes. Some scallops have pretty blue ones. They can live for 20 years. The king scallop is a hermaphrodite with an ovary and testes. Scallop propulsion is also due to their powerful adductor muscle, the muscle that always cramps when I'm running a long race and have to climb over a wall. What is called a scallop on menus should be called 'the scallop's adductor muscle, pan-fried'.

The scallop industry is in the top five most valuable fisheries in the UK.[9] Scallops' popularity has come and gone but now they are so desired that China, Japan, Peru, and Russia farm millions of them. In the ancient world, they were thought to produce diarrhoea because of their excessive sweetness. Baking them avoided loose bowels.

There are two ways to hunt scallops. First, there is hand diving, which is what it sounds like. Divers dive and gather up scallops into a bag using only their hands. It is a slow and precise technique, and hand divers think it is sustainable and the best way to gather scallops. You leave enough so that the next generation can grow; you leave the seabed undisturbed. You pick a fruit from a tree, but you leave the tree intact. But it is dredgers that catch 95 per cent of scallops eaten in the UK.[10] These boats carry six or eight giant metal rakes attached to chain bags. The dredges can weigh two tonnes each, and when they rake along the seabed, pushing scallops into the bags, the teeth can penetrate up to 10 cm. A boat can have up to 22 dredges, each attached to a tow bar that has rubber wheels and is designed to roll the dredge along the seabed and through anything it encounters.[11]

Probably because of the plough-like teeth on their dredgers, scallopers are fond of saying that what they do is no worse than furrowing a field. The sea heals. My accommodation in Tobermory is owned by Paul and Jeannette Gallagher. He is a former scallop dredger and she founded Tobermory's only fish and chip premises, a van on the front. When I stopped Paul Gallagher in the harbour car park and asked him to talk scallops (as his wife had instructed), he said, 'touchy subject'. When we later talk in his living room, he justifies this. 'Well, it's controversial. It wasn't at the time.' His first boat had only one dredge. 'For all you caught you'd probably be better off diving. They used to fill with absolute rubbish: sand and shells. When you lifted the dredge up, you had to break it, the chain bag, and the frame; you had to pull the frame so the chain bag's lying over the side; then you would put the hook into the chain bag and lift that up.' He sees my mystification and explains. A chain bag is what the scallops are dredged into once they have been prised from the seabed. Now there are conveyor belts to alleviate all the triage, but the shells still have to be shucked. Gallagher does not think dredging is damaging. 'The seabed recovers quickly. We fished in this area for years on the west coast and you'd leave an area for maybe a year or so and go back and fish it again, and you'd get a good shot of scallops. Obviously [dredging] kills things, you can't say it doesn't. I would never say it doesn't harm the seabed, but it doesn't damage it beyond, like a nuclear bomb or something like that.'

A 2006 meta-analysis of the 'response and recovery of benthic biota to fishing' found that scallop dredging was the worst fishing method because it wrecked whole ecosystems.[12] It is clear-cutting a forest, not chopping one tree. Slow-growing and long-living animals such as corals took up to eight years to recover from a single scallop dredge. The first study to calculate the carbon cost of industrial fishing, published in 2021, concluded that any gear that disturbs the seabed releases a gigaton of carbon a year.[13] That is more than aviation or Germany produces.

Divers have been telling us about the state of below for years. Davy Stinson is a Yorkshireman now based in Scotland who dives for scallops off Scotland and Norway. He works closely with Open Seas and tells me that the week before, he went diving as usual. 'I wish I'd had my GoPro because there were piles of smashed shells piled up. It was almost picturesque. Like someone had taken out the meat and just piled them up.' Scallop dredgers are inefficient in two ways. 'They kill as much as they land. If you look at a bag of scallops from a dredger, half of them will be full of sand and grit and mud, but you never see it because it's removed in the processing.' And because it is inefficient, they overkill and over-smash, which is why you see tracks criss-crossing. Back and forth and to and fro to get as much as possible. 'I remember going out diving in the late '70s around here,' he says. 'We used to see monkfish all the time. We'd take them for our tea – you'd often say, oh that's a bit small, I'll wait for a bigger one. I don't remember the last time I saw a monkfish. They're not quantified in fisheries statistics.'

Trawling and dredging in Scotland's inshore waters (within three miles of the coast) have been permitted since 1984, when the Inshore Fishing (Scotland) Act was passed.[14] This act repealed a ban on bottom-contact fishing in inshore waters that had existed for 100 years. Since then, Marine Scotland has produced a clunky map with various possible layers, marking in red the areas that are in theory protected from bottom-contact gear. The sea is red from Oban to halfway up the sound to Tobermory. This is the Loch Sunart to the Sound of Jura MPA. In 2019, Open Seas found a scallop dredger illegally operating there.

Not much investigation was needed: the dredger was openly transmitting its AIS data, which indicated that the boat was moving at about three knots an hour, a typical speed for dredging. It was also looping around, another obvious clue. In fact, there was nothing hidden about its operation.[15] Open Seas reported it and there were no consequences. Between 2015 and 2018 Marine Scotland received 22 other reports of illegal dredging in protected areas, carried out by more than 45 vessels.[16] Nothing was done.

Rohan Holt is also old enough to remember what the sea looked like before the trawling ban was lifted, and when Sea Search, a community of diver sentinels, was founded around the same time. He remembers diving in Loch Sunart in the 1980s. 'It has a lovely, very diverse bedrock wall sloping down, then a muddy slope.' He had dived there plenty and knew that once he left the muddy slope, there should be sea pens – animals that look like plants that can grow to 5 ft tall. Rohan remembers 'a forest of them'. The last time he dived there, there was nothing. 'There were linear scar marks on the seabed and all the sea pens had disappeared. Everything had gone.'

In 2017, locals near Loch Carron reported sighting a scallop dredger working in the loch.[17] Some of the locals knew that Loch Carron contained fragile flame shell beds. Flame shells sound inanimate but they are animals related to clams. They live on the seabed in nests that they build from marine scatter such as shells, rocks, and whatever suits them. 'Hundreds of nests,' writes Nature Scotland, 'can combine to form a dense bed, which raises and stabilizes the seabed and makes it more attractive to many other creatures.'[18] Flame shells are important and precious, or in the words of the Encyclopaedia of Conservation, they are 'a protected keystone biogenic bivalve habitat'.[19] They are also, says Rohan, 'incredibly fragile to anything that's dragged across the seabed'. I ask if there is anything that isn't incredibly fragile to anything dragged across the seabed, and he says, 'Mobile sand. If you're in an area where there is a lot of wave action, maybe a fair bit of tide as well, but the seabed is always kept turning, big wave action. So the sand is kind of very loose and you can put a dredge through that and five minutes later you wouldn't see any signs of it. You wouldn't do it anyway because scallops don't like it.'

After the dredger had been sighted, some divers went diving. They had cameras and recorders and they recorded loss devastating and shocking enough to make the BBC news. The flame shell reef was smashed. There were clear tracks gouged across the seabed and smashed red shells everywhere.

The outcry triggered the Scottish government into declaring half of Loch Carron an emergency MPA, then a permanent one. Open Seas wrote, 'What happened in Loch Carron is a window into the bright future that could exist along our coastline.'[20] During a 'celebration dive' that year, the diver activists who had recorded the devastation found recovery: 'Undamaged and healthy sections of flame shells next to tow marks have been able to partially recolonize the dredger's tracks.' The flame shells were thinner than the ones in non-dredged areas, but they were there.

On the small island of Seil, near Oban, David and Jean Ainsley run dive charters. Jean is an accountant; David is a full-time diver and activist. Along with Davy Stinson, the Ainsleys have been the sentinels of their local undersea by diving and watching and filming and following it up with court cases and legislation and endless meetings with Marine Scotland. We talk while sitting on their anchored boat on a lovely afternoon. David has been a dive-detective for more than ten years. 'It was in about 2004 that we started a campaign to stop scallop dredging. So I would go and film dredged seabeds. I filmed one bit where dredges had run across a rocky reef and there was unimpacted rocky reef on one side. And then we swung the camera across to the rocky reef where all of the soft corals had been pulled off by dredges. You could see all the scratch marks across. And the damage is very obvious.' A natural and horribly unnatural before-and-after. At first they tried the good citizens' route, approaching a local marine protection committee. 'But we were always voted down.' So then they went on TV, onto a well-known Scottish nature programme called *Land-*

ward. 'In those days I think everybody saw fishing boats going up and down and, you know, it was very much, here are our brave boys out at sea. And nobody really realized the damage that mobile gear fishing is doing. When it was on *Landward*, it was quite an eye-opener for a lot of people that actually these boats are tearing up the seabed and destroying important habitats.'

The usual arguments from scallop dredgers fell on ground as bare as the seabed they were leaving behind their dredges. 'You know,' says David, 'they say, "we are tilling the soil". Well, first of all, you don't own the soil. And if one particular type of fishing is destroying that commonly owned resource, then that is unlawful.'

All the Ainsleys' work was unfunded. 'The other thing is they always say that all these environmentalists are funded and all the rest of it, and I wish we were.' Diver activism isn't cheap: there is the running of the boat, the gas, the film, the equipment. 'We're up against businesses who make lots of money, who pay lawyers. And we are just ordinary people who can see that the laws protecting the environment desperately need to be enforced.'

Still, they kept diving. This was the good news period. They could see signs of recovery. It was nothing yet like the seabed in the No Take Zones they had dived in off New Zealand and Indonesia. In their local waters, there had been skate and spur dogs everywhere before the dredgers came and they hadn't come back. 'Even so,' says David, 'the recovery has been really quite dramatic'. Drama is easy when the bar is set so low. They also kept an eye on the sea above. In November, David noticed a scallop dredger illegally fishing off Colonsay. He remembers the month because scallop prices go up at Christmas. 'I have to admit that you couldn't be absolutely certain that you could see his warps in the water from the photographs. But his VMS has been disabled.' (A VMS, or vessel monitoring system, transmits a vessel's position but the data is encrypted and usually reserved for governments and fishing authorities.) 'We know that that's what they do. They put a metal bucket over the thing on the roof. But we knew he was out there.' David's photographs were good enough to clearly identify the boat and he sent them as evidence to Marine Scotland. 'They told me the law was that he was allowed to fish for 24 hours without his VMS working.' This is not what Marine Scotland's website thinks: according to its marine and fisheries compliance vessel monitoring system page, 'UK vessels are allowed to power down in port only.'

Then Jean saw another dredger off the Garvellachs, a group of islands to the south of Seil. The boat was dark: they could only see the wheelhouse light. The boat was obviously fishing illegally: it was dark and in a place known for scallops. David phoned the Marine Scotland compliance line and reported the boat's position and asked them to investigate. Nothing happened. The next night, the boat was there again. I remember a cartoon in my childhood called *Hong Kong Phooey*, in which a beagle who works as a 'mild-mannered' janitor frequently turns into a crime-fighting hero called Hong Kong Phooey. I think of this because 'mild-mannered' is a fair description for David Ainsley, but he must have gone Phooey, because he called in the troops, in the form of his friend Dirk, who had a scallop diving boat.

They set off to the Garvellachs, six miles off, equipped with a bag of cameras. 'As we approach, the behaviour of the boat changes. All the lights come on, and he starts steaming up towards Oban. We come up alongside and we film the boat with his dredges kind of badly stowed, hurriedly stowed, and huge bags which we presume contain scallops on the back of the boat.' They got it all on film. They got the dredges, the boat number, they reported it once more to Marine Scotland, and finally someone came to take a statement. 'They told us the procurator fiscal [the local prosecutor] won't prosecute because we didn't see the boat operating with the dredges in the water. Now, let's say this was a murder case, and I'd been a witness. I'd filmed the man running away with the knife covered in blood. But because I didn't actually film the knife going into the victim, would it be thrown out of court?'

After that, he marshalled the divers. The local scallop divers, the local sports divers: everyone came out and they searched. They knew where to look because they had lined up the boat's wheelhouse lights with the island. 'We filmed very freshly dredged scallop ground. We had evidence of the crime on the seabed. You couldn't have better evidence.' Still Marine Scotland did nothing. 'So that really convinced us that [they] just don't want to enforce the laws protecting MPAs.' But the BBC covered it, and there was a big meeting of locals, scallop divers, creel fishermen, recreational divers. 'Just people who wanted the seas to recover.' It was probably, he thinks, the seabed from which Open Seas grew. 'An awful lot of good happened,' he says, 'and I haven't seen illegal dredging since.'

Day two on *Sea Beaver* and the air of frustration does not match the serene surroundings. The ROV is still not working. I note down surrounding boat names to pass the time. *Snifter. Favourite Child.* A jellyfish appears off the pontoon and Theo arrives with his knowledge. 'That's a quarter-grown lion's mane, they grow up to 30 metres, they live for a year. This one has lost its tentacles.' Neil says afterward, 'Has Theo been giving you his jellyfish facts?' I don't mind Theo's enthusiasm: it is a gust in the doldrums.

I have been hanging around for a couple of hours feeling stoic because I will not get upset about this not working nor think of the money this trip has cost me. This is not lavishly funded military science. It is not lavishly funded anything. It is a small organization trying to gather robust data with unsatisfactory gear. The ROV cost £6,000 ($7,500) and that is the cheap end of the market.

I dive back in with Rohan. If I can't see contemporary investigation – beyond Theo swearing at the instruction manual of the ROV – I want to know about the previous five months of the survey. I ask if *Sea Beaver* has gone 'Sea Shepherdy'. The ocean protection organization Sea Shepherd does excellent investigation, but it is also better known for loud actions rather than quiet and patient ROV research. No, says Rohan. That's not us. 'The aim isn't to point fingers at any individual but it's to say to government, your actions aren't working, this is what people are getting away with. We're not interested in prosecuting individuals because it doesn't actually help anything – it's too late, that's done. Government is supposed to look after the fish stocks and the seabed. So we've been backwards and forwards chasing people with drones and stuff.' Shouldn't Marine Scotland be doing this? 'We've seen a fishery protection vessel twice in the last four and a half months.' In north Wales, where Rohan is from, 'the marine protection's fuel budget allows them to go from Conwy to the Menai Bridge, which is about 15 miles, then it runs out and they can't go any further than that, so they just do that now and again and go backwards and forwards. It's like nothing at all.'

Rohan tells me the tale of an offender whom they caught in the act. It was 'a shonky horrible boat crewed by a bunch of cowboys going into MPAs, mopping up bits where they shouldn't be, because they know the profits are pretty high because no one else does it'.

One day, *Sea Beaver* had been 'basically bimbling about' around Wester Ross, with the ROV (a working one) operating about 7 metres below the surface. Rohan is genuinely excited now and indicates the video now playing on his laptop. 'And *that* swims past. A flapper skate. It's a critically endangered species and I've never seen one before in the wild.' The *Sea Beaver* crew were ecstatic. 'Everyone was going, yay, look at this. This thing flew past and we managed to keep up with it with the ROV for a little bit. Really quite a large animal. The biggest one recorded had a wingspan of nearly three metres – they're huge. We surprised each other in a big way.' In the UK, it is illegal to land a skate. If you catch one, it must be released immediately. The crew had investigated various nets in various ports and had once seen two dead baby skates in a net.

Then on another day their drone was hovering above a trawler. 'We were looking for undersized catch, just to see what they were doing. So there's thousands of little fish here, all of them undersized, then there's this thing on the deck. That is a flapper skate, you see the size of it? You see the two nets coming in, they get dragged up onto the back. You see the guy starts to struggle opening one of the nets to release the contents onto the back deck, and as he does this huge flapper skate just goes blumph.' (He means the skate dropped from the net to the deck.) The crew should then have immediately dropped it back into the water. Instead, they passed a rope through its gills and mouth and hung it up, and the boat steamed to the fishing port of Peterhead non-stop, another indication that the skate was illegally landed and illegally sold, probably for export.

The skate was one crime. The undersized fish were another. And the throwing of other dead fish into the water was yet another. All this on 10 minutes of drone footage. Just one boat, in one small part of the sea, doing great wrong.

The replacement ROV is finally assembled. It needs to be tested, so the boat steams out to a small island opposite Mull. It is agreeable: water like glass, a small islet covered in seaweed, seals sunbathing and swimming. A man in a wee yellow boat arrives, dumps two creels, and departs at speed. Peace and beauty on the third day of Scotland's three-day summer.

The ROV sets off. It descends as it should. The screen transmitting its images shows murky green. Where is the wall? Head for the wall. The ROV comes back with its battery showing as flat. Kim revs the engine every now and then to get the boat back in the right spot. Drift then rev, drift then rev. Rohan says, 'If you want civil-led science then this is what it's like. Proper kit for subsea surveys is a hundred grand [$124,000]. And otherwise it is literally fixing it yourself with gaffer tape.'

The gaffer tape is no use. The battery is no use. 'You have been here on the two quietest days of the whole trip,' says Kim. We steam back in sadness. The next day the crew say they are off to the far side of Mull, and I make my way back to Oban via bus and ferry.

The Scottish shellfish industry is represented by the Scottish White Fish Producers Association (SWFPA). It has a scallop committee and powerful tentacles because the shellfish fishery is worth more than any other in Scottish waters. The SWFPA has said that illegal dredging is 'entirely associated with a few rogue vessels'.[21] In 2021, a scalloper was fined £187,170 ($238,330) by a court in Scarborough for illegal dredging offences, including operating in a closed season; using scallop dredges without the authority of a permit; exceeding the permitted number of dredges; failing to operate a fully functioning vessel identification system; and landing undersized scallops. The boat belonged to John MacAlister, former chair of the SWFPA scallop committee.

The marine ecologist Daniel Pauly published an essay called 'Aquacalypse Now' in 2009.[22] It is refreshingly furious. Nothing in it is irrelevant because nothing much has improved since then. Our oceans, he wrote, have been victims of a 'giant Ponzi scheme . . . This scheme was carried out by nothing less than a fishing-industrial complex: an alliance of corporate fishing fleets, lobbyists, parliamentary representatives and fisheries economists.' These actors hide behind 'the romantic image of the small-scale, independent fisherman'.

The Scottish Creel Fisherman's Federation, an ally in the environmental activists' battle against dredgers and trawling, wants the reinstatement of the three-mile limit around Scotland.[23] Back to 1869. Or back to 2008, when a 1.7-square-mile area of Lamlash Bay off Arran was declared a No Take Zone. 'No Take' means what it sounds like. No fishing, no machinery, no creeling, no pots, no nets. In the words of the Marine Protection Atlas, a No Take Zone 'prohibits the extraction or destruction of natural or cultural resources'.[24] After a decade, scientists from the University of York tracking its biodiversity found that there were four times as many king scallops as in 2010. The scallops were bigger and there were more juveniles. There were also more lobsters, and 'the growth of structurally complex "nursery habitats" which provide refuge for marine life'. If only 'protected' in every 'marine protected area' could solidify like a clam shell, but faster.

Crew member poses with the head of a dolphin on board a Chinese longliner authorised in the South West Indian Ocean

11

A Dedicated Fish Warden

Imagine a terrible job. Probably you will be thinking of filth and danger, because of all the TV series about crime scene cleaners and sewer workers and ice-road truckers. The hardest job in the world does include filth and danger, but it does not appear on TV or in everyday culture. This job requires you to spend several months with a possibly hostile group of strangers who don't want you there, in a situation where killing you is as easy as a push. It requires you to eat with them, share accommodation with them. You will have no privacy. You will probably have no independent communication, no emergency transponder, no radio, no internet, no phone. You are not employed by the people you have to live among but you are paid to report on them. If anything happens to you, you will usually be several days from safety. You are a fisheries observer.

On 4 March 2020, my calendar tells me I took a yoga class, had a therapy session, took my cat to the vet, and talked to a woman who was putting on an art show about container shipping. A busy day. On 4 March 2020, Eritara Aati Kaierua was found dead in his cabin on a tuna boat. Eritara, 40 years old and a father of four children, was a fisheries observer from Kiribati, an island in Micronesia. This means he was employed by the Kiribati government to work on tuna boats and collect data on the catch and by-catch, among other things. In the words of Human Rights at Sea, a campaigning charity, 'An observer is a person who is authorised by a regulatory authority to collect biological and operational data from commercial fisheries.'

Eritara had spent 12 years working on cargo ships. By then, he had a young family and he wanted to be home more often. Observing meant weeks away, not months. In 2020, he was in his eighth year of observing, and his usual job was tuna boats, because that is the main business of Kiribati's nearby oceans and of the western central Pacific in general. In February, he was assigned to the *Win Far No 636*, a vessel flagged to Taiwan and owned by a Taiwanese company. It is obvious that a name like that will be Taiwanese or Chinese, as they produce many vessels with the same name – like a car production line for ships – and distinguish them by number. In 2016, wrote Human Rights at Sea in a report, 'Eritara allegedly told his sister about attempts to bribe him over a shark fin catch. In 2019, a crew turned on him when they were forced to offload tonnes of tuna in Tuvalu after officials found his log did not match that of the captain.'[1] He was so scared, said his sister Nicky, he thought he would be poisoned. 'Most of the time he would eat noodles and biscuits, his own rations, in his room. He came off that boat and he reported it to Kiribati Fisheries.' His employers' response was to assign him to that boat's sister boat. Same owner and operator. 'That showed,' said Nicky, 'that safety wasn't a priority. But he came back alive, and he was really thankful for that.'

What kind of job makes you glad you make it home alive?

On his last trip, Eritara boarded the *Win Far No 636* at 2.20 p.m. on 13 February 2020. It is impossible to know where the boat then went because its AIS was deactivated. In his last email, Eritara had worried about his wife's health and encouraged his children to 'go hard at school so that they become intelligent and wise'. The RFMO Western Central and Pacific Fisheries Commission (WCPFC) has minimum standards for observers. They include the requirement for an observer to have his or her own two-way communications device. Eritara didn't have any such thing, even though New Zealand had donated AUS$26,000 (approximately US$16,000) worth of two-way devices to Kiribati.

A Dedicated Fish Warden

They never reached the observers.[2] He had to send his message from the boat's email address, with no privacy.

He signed off by saying he had to go because 'the fishing net is going to be [set]'. He did not mention any problems or issues with the crew and 11 days later he was dead. On 6 March, just after midnight, the crew of *Win Far No 636* reported Eritara's death to the WCPFC and to the Kiribati Observer Coordinator. WCPFC safety measures required the boat to immediately stop fishing and to head for the nearest port for investigation, which was Nauru. Instead, the boat made for Kiribati and arrived a day later than the trip necessitated. The reason given was bad weather. Eritara's sister Nicky, in a documentary about his death, was sceptical. 'We've had access to the details of the weather, and we got the details of their speed. They were just sitting out there for hours. What were they doing? Were they cleaning up?'

Twenty days later, an autopsy performed by a Fijian pathologist found that Eritara's death had been caused by 'severe intra-cranial haemorrhage . . . and traumatic brain injuries due to severe traumatic head injuries and blunt force head trauma'.[3]

I encounter Eritara's widow, Tekarara, and Nicky in human rights reports and in a documentary by Sara Pipernos called *Death at Sea*. A former fisheries researcher, Pipernos switched to making films because she thought it was a more powerful way to change things than writing yet more reports. Pipernos intended to focus on the disappearance of another observer (she had dozens of cases to choose from) but then found Nicky and Tekarara, 'two insanely impressive women', and Eritara.

Eritara's sister and widow asked plenty of questions, of everyone they could think of. It was quite a list, as Eritara was employed by the Kiribati Ministry of Fisheries on a Taiwanese fishing vessel but fishing in Nauru waters. The women got no answers. Two subsequent autopsy reports were commissioned, but Nicky and Tekarara suspect they were ordered by the Taiwanese fishing company, and they were carried out remotely by pathologists who had no access to Eritara's corpse. Both concluded that Eritara had died of hypertension. Natural causes.

'Initially my brother didn't die of hypertension,' says Nicky on camera. 'Seven months later he died of hypertension.' She says this cause of death has been attributed to other mysterious deaths of observers. 'There's a number of them dying of hypertension. And I'm wondering if it's an easy way out.'

In a judicial review of the case, I find a photograph of the crime scene. It shows Eritara lying with his head thrown back. His T-shirt is vivid green; his shorts are striped; there is blood on his pillow. It does not look natural.

Search for 'fisheries observer' and 'disappearance', and you are most likely to find reports of Keith Davis. He was an American who worked for MRAG, a private agency that supplies fisheries observers, and he often worked on trans-shipment vessels. These are larger boats that accept catch from smaller boats at sea, provide bait and supplies, and sometimes move crews around between boats. Trans-shipment means that boats and crews are obliged to stay longer at sea. Observers know that trans-shipment jobs are the riskiest, because trans-shipment is where the worst things happen because they can: illegal catches of sharks for shark fin, contraband hidden among the fish, fishes that are not the fish the label will say it is. Every second, according to the Pew Trusts, more than 800 kg of fish is caught illegally.[4] Most of this happens in international waters. Some of it happens on boats with observers on board.

In early 2015, Davis and two co-editors published a book written by fisheries observers called *Eyes on the Seas*. In September that year, he disappeared from the *Victoria 168*, a trans-shipment vessel that flew the flag of Panama and had a crew of Chinese and Burmese and a Taiwanese skipper. No trace of Davis was ever found. His deck wear and survival suit were in his cabin. The investigative journalist Sarah Tory, writing for *Hakai Magazine*, concluded that Davis had probably noticed evidence of human trafficking or migrant transportation.[5] Weights that could weigh a body down in water were missing from the gym rack.

Alfred 'Bubba' Cook is a tuna expert at WWF and on the board of the Association for Professional Observers (APO). He refers to Davis's death as murder. When I query this word, he says, 'That's my opinion, and I have absolute certainty in that assessment because there's no way that he would've accidentally gone missing from that ship. And if you've done the research, the entirety of the circumstances involved with his disappearance, they can lead to only one conclusion.'

The APO attempts to keep a list of suspicious deaths and disappearances, painstakingly gathered from news reports, tweets, and blog posts. Liz Mitchell-Rachin, APO president, has recorded one or two deaths of observers every year since 2015, but she knows that they don't know if that is an accurate number.[6] What is clear is that there is a pattern to what happens post-mortem, whether the observer is a well-connected American citizen or someone from an island nation with the lowest GDP in the Central Pacific. 'All cases have the same outcome,' says Mitchell-Rachin. 'No information. The playbook of each disappearance or death is the same. The news leaks out from unofficial sources. The families are told, "We're looking into it." They hear nothing. They express grief and anger, and they are overwhelmed with questions.'[7]

This is not about the poor capacity of developing nations to mount investigations. Rich ones aren't good at that either. Because Keith Davis was a US citizen, the US Coast Guard and FBI flew to Panama to investigate, along with two MRAG officials. They were not allowed to direct the investigation, because Panama was the flag state and had control. I once wrote about how Panama for years refused to release an investigation into the sinking of the livestock transport ship *Danny FII*, just because it could. Panama has a history of obstructing and procrastinating. The FBI was willing, says Graeme Parkes of MRAG, but they were hampered by knowing little about the world they were investigating. 'What are they doing again? Fish?' The US Coast Guard Investigative Branch was also involved. 'They do know fishing,' says Bubba Cook, 'but the Panamanian authorities basically locked everything down. They didn't allow the US law enforcement agencies to really do the work they were entitled to do.' Of all the disappearances and deaths listed by APO, hardly any have involved a satisfactory investigation.

David Hammond, a former lawyer who runs Human Rights at Sea International, is diplomatic and also frustrated. 'Kiribati has put itself in a position that doesn't look good.' Pages of Eritara's logbook went missing. When his widow visited the Ministry of Fisheries to ask questions, she was told her husband's possessions were with the police. No, she said, pointing across the office. That's his wristwatch right there. And that's his notebook. When she asked to see her husband's contract, she was advised to hire a lawyer. 'A lot of the mobile phones taken from the crew have gone missing,' says Hammond. Even though the Kiribati authorities initially sought help from Interpol and the US Coast Guard offered to assist, nothing came of either, and Eritara's family don't know why. 'It's deeply frustrating,' says Hammond, 'but to be honest, it goes with the territory.'

Four years after Eritara's death, Kiribati police have not closed the case. The *Win Far No 636* has long since been released and is back fishing. Tekarara is yet to be given the workers' compensation she is entitled to, as well as a conclusive answer to why her healthy husband suddenly died. In his final email, Eritara, a 'humble and happy man', worried about his wife's illness. She should make sure to take her medicine, he wrote, so that she recovered. 'I will try my best to stay healthy from here too.'

Fisheries observers are relatively new on the seas. There have been scientists aboard vessels for centuries, including Charles Darwin. The concept of an observer – an independent presence who would monitor and report back – was different. In the United States, observers began to be placed on boats in the early 1970s. In 1976, the United States had enacted the Fisheries Conservation and Management Act. Nicknamed the 'Two-Hundred Mile Bill', the Act gave US fishermen first preference over stocks within the 200-mile limit. Foreign vessels could still fish, but now they had to carry observers if the US government required it.

Before observers, fisheries managers wanting to understand what boats were catching and what they discarded had to rely on honesty and on what skippers recorded or told them. Or they watched from docks and wharves as the catch was landed, but this gave them no measure of what had been discarded. Observers were unique: they were there. And they were supposedly independent. Soon fishing crews gave them a name. 'Feds' in US fisheries. 'Fish cops' or 'spies' everywhere else.

Fifty years on, observers can be deployed and employed by private agencies such as MRAG or Archipelago. They can be employed by a nation state. They can work directly for RFMOs or for the Inter-American Tropical Tuna Commission trans-shipment observer programme.

Today the observer's job varies as wildly as the employers do. In *Eyes on the Seas*, the three editors – all working observers – include an incomplete list of what they have been tasked to monitor. 'Closures of Marine Protected Areas (MPAs); quota restrictions (limits on the catch of target species); incidental take statements (ITS), which identify limits of certain incidentally caught protected species and by-catch, or both; permit restrictions limits on how many vessels, what gear types, and where and when fishing vessels are allowed to take marine biological resources; technical measures, which define the permitted technical characteristics of the vessels and gear deployed in the fishery; or, a combination of such measures.'[8] Whatever they do, they are, in the words of Greenpeace, 'some of the most isolated human rights defenders in the world'.[9]

Online, there is no shortage of young people eager to be fisheries observers. It is a popular choice for biology graduates who think it will be useful on a CV. It is seen as adventurous. Most people applying to NOAA's North Pacific Observer Program come from landlocked states. There is a pattern on online forums. An anxious young person asks whether the job is a good idea, and the answers take two forms: it can be hard but you'll love it, or it can be hell and for god's sake look for a different job. The scales usually drop on the negative side but not always. 'Depending on where you'll be,' says one observer, 'it can be hard but not as hard as the fishermen working. Back in the day, fishery observers were threatened and abused. Those stories are more few and far between these days. You count fish and record data. Pretty easy.'[10]

If it's that easy, why don't people stay in the job for years? Probably because even with a fair wind and a good crew, the lifestyle is arduous. You're away for months and it is impossible to plan because you are bound by the habits of fishes. The work is hard: you work when the crew work, and they work a lot. You will smell of fish no matter what your efforts at cleaning. You will get used to sleeping through engine noise and motion (take a good pillow) and snores.

You will probably have a bed and sometimes your own room, but often you'll have to share. Sometimes you have to hot-bunk: you share your bunk with a crew member and sleep when he works, and vice versa. Sometimes your bunkmates are not human. Observer Chris Stump: 'On my first vessel [the roaches] were so bad you had to line your bunk with duct tape on that boat, sticky side out to try and establish a barricade while you were sleeping . . . If you were on a bedbug-infested boat, there was nothing you could do. You couldn't exterminate them by just killing the ones you saw, and you couldn't move away from them because they were in everything (made of fabric) that you owned. You just had to sit there, let yourself be bitten and hope for a short trip.'[11] Who and what isn't a potential problem when you are an observer at sea? Observer Ethan Brown's summary of his job: 'an endless wet nightmare of huge seas, puking, pork chops, and sexing crab'.[12]

Your obligations may change, but usually you will have to estimate the catch, quickly distinguishing between almost identical species because one has catch limits and the near-identical one doesn't; you will measure fish and remove their otoliths to assess age and growth. If the catch brings in 40 species of fish, you have to identify each one. You do this in every type of weather, sometimes while you are lashed to a rail. You'll often be covered in fish guts and fish sperm. You must be alert to the dangers around you, which on a fishing boat on the high seas are manifold. You should keep your head on a swivel. 'My training,' said former Alaska groundfish observer Natalie Posdaljian, 'should be called 1,001 Ways to Die. Everything on a fishing boat . . . you're constantly just watching out for something that could kill you.'[13]

You may also be tasked to record what is being discarded and how, if you're allowed to. When *Vice* spoke to four Canadian women who had worked as observers, none would give her real name although none was still working as an observer. All had been employed by the private observers consultancy Archipelago and all had horror stories. First the fishes: most trawlers net by-catch but there are ways to minimize harm. Ideally any non-target species are thrown back overboard alive and with a chance of survival. Instead, on one factory freezer trawler, 'Erin' watched as crew members stabbed sharks in the head because they could. 'Birds that landed on the ship . . . were stomped on and crushed. Skates . . . were torn in half by a crew member while he looked directly at me.'[14] On one forum thread, someone makes a point that I'm surprised is not made more often: 'Being an animal lover, it was difficult to see so many fish and birds killed every day, but I also got to see whales, dolphins, sea turtles, flying fish.' One observer who worked in the 1990s found fishing crews were always cooperative. They weren't the problem. 'I preferred midwater trawl boats because they didn't have a lot of by-catch. Dragger boats (bottom trawlers) were depressing though. So much by-catch, debris, and trash just dumped back into the ocean.'[15]

A comment from a user called Wazeppelin: 'You are much more at risk from the vessel/sea itself than you are from a fisherman. They will go to extraordinary lengths to save you if you were in danger and you were the worst observer ever just because that's what you do for anyone at sea.'[16] That's probably true. Onboard behaviour can be less chivalrous. When the NGO Women in Ocean Science surveyed 980 women working in marine science, it found that 78 per cent had reported being sexually harassed. Most commonly it happened at college or in the field, and sometimes the field was the sea. Even when they had access to help, 39 per cent of women did not report it.[17] When NOAA conducted an anonymous online survey of the 350 observers deployed in Alaska (half are female), a quarter of female observers responded, and half of those said they had been sexually harassed at sea. Hardly any had reported it.[18] Another survey found that only 20 per cent of observers felt valued by the fishing industry.[19]

Jenn Ferdinand is responsible for the North Pacific Observer Program. She used to do the job and never had a problem. Women now get no special treatment. 'We don't place observers on boats based on their gender identity because that has been an issue in the past. We've had boats that said, we don't want a female observer. So now we say you're going to get who you're going to get.' And who will the women get? Almost certainly a boat full of men. Women are much more present in observing than fishing, at least in the US.

Under the US's Magnuson-Stevens Act, it is unlawful to 'forcibly assault, resist, oppose, impede, intimidate, sexually harass, bribe, or interfere with any observer on a vessel'.[20] Transgressors can get ten years in prison, a fine of $200,000, or both. APO laments the lack of reporting of harassment from observers in regional management programmes. It has heard 'directly or indirectly' from observers that they have been locked in their rooms, threatened at knifepoint, chased down the dock, forced to sign off on MSC criteria (the MSC sets standards for sustainably caught and traceable seafood), forced to accept a bribe under threats of violence, raped, had food withheld.[21]

In the world of fisheries observing, NOAA's North Pacific programme is considered to be unparalleled. The programme is huge: it covers the Bering Sea, Aleutian Islands, and Gulf of Alaska groundfish and halibut fisheries. It is comprehensive: in 2022, observers covered nearly half of all fishing trips.[22] Worldwide, most RFMOs and nation states aim for a rate of 5 per cent coverage of fishing vessels and often don't manage that.[23]

NOAA's programme is funded by a mixture of congressional allocations (the minority) and fees paid by the fishing industry (the majority). It employs nearly half of US observers: in 2019, 850 observers monitored 54 fisheries and spent 71,600 days at sea. 'We have really big boats that can accommodate 140-plus processors,' says Ferdinand,

'and we have tiny little boats that just hold three or four fishermen. And we observe the whole lot.' Observers can be on draggers, potters, or long liners. Groundfish can be pollock, sablefish, Pacific ocean perch, or sharks.

Observers have fisheries favourites. Alaska is good but not the most lucrative. A Hawaii job will put you on a vessel with Japanese, Micronesians, or Koreans, and it may be lonely. 'Observers on Hawaii boats really make bank,' says Natalie Posdaljian. 'That's because they're working with people who don't speak English, which isn't such a big deal. But you might be out at sea for a month and not talk to anybody for a month.'[24]

Bubba Cook of WWF is convinced that observers are an essential tool and that any fisheries scientist should be fighting their corner with ferocity. 'We know we can't trust the fishing industry to give us accurate data. And we know that because when we have observers on board, we very often get very different data than we get from vessels that don't have observers on board.'

IUU could also stand for illegal, unreported, and unregulated human trafficking and labour. Trafficking is the most profitable transnational criminal enterprise, and the unpoliced high seas is a very good place to do it. Obviously the cure for dark behaviour is light and transparency. Who is meant to shine that light? Fisheries observers disappear, and seafarers disappear more frequently. A maritime emergency call may be responded to in days not hours, given the size of the sea, or it may not. People who talk of lawless oceans are wrong. There are countless laws but there is no one to enforce them. The much-praised Port State Measures Agreement (PSMA) is supposed to eradicate IUU fishing by preventing IUU vessels from landing their catch.[25] It's a good initiative. The PSMA is little use when you are 100 miles off the Galapagos and dying from beriberi because your diet is instant noodles and fish bait and you have been at sea for a year because your catch is constantly trans-shipped far from any port or any measures.

The Indian Ocean Tuna Commission keeps a list of IUU vessels. It is currently 12 pages long. It has plenty of detail: latitude, longitude of vessels, sometimes AIS, sometimes IMO number. This is a comprehensive dataset of wilful wrongdoing. In most instances in the 'action taken' box, it reads UNK/INC – unknown and inconnu.[26]

What observers give us, says Cook, is 'reality-based fisheries management'. What's the opposite of that? The opposite is what we have, which is guesswork. 'When you only have 5 per cent observer coverage, then the information that you're working from is statistically insignificant. The amount of information that you're getting from the 5 per cent of observers serving on board is really just a thumb in the wind. It's a best guess.' Most calculations of fish stock sizes are based on incomplete data and projection. 'I tell my fisheries scientist friends that they're polishing a turd and they're doing it very well – they're very bright people – but it's still a turd.' Observers, says Cook, are never not going to be essential. He says at least 20 per cent coverage is ideal, but it is a dream.

It is not the job of a fisheries observer to report on human rights or labour violations. Sometimes they do, if they dare. They are called the 'eyes on the sea', but when the average observer coverage in fishing fleets is so low, there can never be enough eyes. Luckily they are not the only ones watching out.

On the north-west coast of Cebu Island in the Philippines, facing the Tañon Strait, there is a small town called San Remigio. It is a town of fish people. They look outward to the largest MPA in the Philippines, a strait that is a 'favourable cetacean habitat'.[27] It is also one of the most heavily fished areas in the nation. Tañon Strait, writes Oceana Philippines, is 'rife with pirate fishers' who target squid, sardines, and of course tuna.[28] Everyone always wants tuna. (It is the Philippines' most exported seafood product after seaweed.[29]) The pirate fishers are illegal trawlers and also smaller boats that use dynamite. This is illegal in 'municipal waters', a stretch of 9 miles out from shore restricted to fishing boats weighing less than 3.1 tonnes that use passive gear such as hook and line or gill nets.

A Dedicated Fish Warden

Norlan Pagal, from San Remigio, is fisherfolk. He began fishing as a child, accompanying his father. 'I remember then that the sea was abundant with fish.' They fished for mackerel and tuna using hooks and lines, and also octopus, using a technique that the ancient fish-lover Oppian would have liked: they dropped a stone wrapped in black cloth in the water, and the octopus chased it. It was subsistence fishing mostly, but sometimes they sold fish for profit. Norlan is now 52 years old, and the waters he knows so well have changed. 'Before if we fished for 3 hours, we could catch 20 kg of fish. Now if we fish for 12 hours, we can't even catch 10 kg.' In a publication entitled *Love Letter to the Tañon Strait*, Oceana wrote, 'Fisherfolk [catch] roughly 90 percent less for the same amount of effort as they did in the 1950s.'[30]

By 2022, this visible decline pushed him to do something. It sounds like something simple, unimportant: 'I spoke to my co-fishermen, I told them that we need to protect our seas, and to get rid of illegal fishing.' So they created a local *bantay dagat*, a community-led fisheries patrol.[31] The first *bantay dagat* ('Guardians of the Sea') had been founded in 1994 by Senator Santanina Rasul, who had been horrified to see 30 fishing vessels operating in protected waters.[32] Supported by the Philippine Navy and Coast Guard, each unit of *bantay dagat* was meant to be a community-based force that carried out foot and sea patrols using speedboats and local outrigger boats called *pancas*.

Norlan and his co-fishermen began with foot patrols then took to the sea. For the next 17 years he saw everything that he should not have. Fishermen using dynamite to blast fish from the water. Fishermen diving down using compressors – breathing through narrow hoses, a spectacularly precarious way to dive – or mixing cyanide with minced fish meat to serve as bait. He saw foreign vessels but also people he knew. The illegal fishers of the Tañon Strait are not usually foreigners. Sometimes the illegals were his relatives. In 2010, he was in his boat at sea when he confronted a boat using dynamite in the strait. They threw the explosive at him. Eight people; eight sticks of dynamite. Most landed in the water. 'But one stick landed on my boat. By the grace of God, I was able to brush it off with my hand; it exploded in the water instead.' He knew the men had come from the Bantayan Islands, but no one was arrested. 'We had begun patrolling at 5.45 a.m.; the Philippine National Police arrived at 10 a.m. But the illegal fishers had already escaped.'

Three years later, he arrested some illegal fishermen, who then cut his head open with an oar. He describes both these incidents in one video as 'memorable experiences'. I ask how his wife and children felt about this. 'At first they feared for my life. But then they understood that my advocacy was for our community, and they supported me.'

In 2006, a local newspaper reported that there was a bounty of one million pesos on Jojo de la Victoria, head of a *bantay dagat* that was trying to protect the Visayan Sea north of Tañon Strait.[33] Two days later, a serving police officer shot de la Victoria dead. Jojo de la Victoria – described in the national senate as 'a dedicated fish warden'[34] – had been Norlan's mentor. 'When he died, I became more persistent in protecting the seas, because I knew a life had been lost.' Surely you were scared, Norlan? 'No. My fear was entirely gone. Who would stand up and protect our seas if I let my fear win?'

Three years later, Santiago Dulay was shot in Nasugbu town, where he had been head of the *bantay dagat*. Then in October 2015, it was Norlan's turn. He had been celebrating the yearly fiesta in his village of Anapog and was heading home to change when he noticed two men on his street. He walked past them, got to within 30 metres of his house, then turned back in time to see something glimmer. It was the gun. 'I was shot only once,' says Norlan. Only. He pretended to be dead and they left. He could not shout. 'My voice was low because I was shot.' His mother had a shop nearby: she came to him.

His spinal cord was severed, and he was paralysed. No one has ever been tried for the attack, although Norlan knows who did it. 'Illegal fishers paid for it.' He recognized the assassin from seeing him at sea. He was not even surprised to be shot, because a friend had warned him that someone had been paid to kill him. That friend was one of the two men.

'What?'

'Yes, my friend was also a gunman.'

One of the gunmen escaped to Manila. The man who paid for the murder, who owned compressor fishing vessels, died of natural causes. I ask if Norlan knew the man who wanted him dead. 'Yes, I know him very well. He's from San Remigio.'

Norlan and I speak over Zoom. He has come to the local village hall to get a signal, and the signal disappears a couple of times. Technology is not relevant to what Norlan does: he cannot rely on it. His surveillance is analogue. Even or perhaps especially after he was shot, he still went out to the shoreline in his wheelchair, just to watch out. I ask if his wheelchair is a good one. 'Not really.' He was given one by the former vice president of the Philippines but that broke. The municipality gave him this one, 'but it's only really good for temporary use'. He speaks in Cebuano, but sometimes uses English words. These include 'marine protected area', 'advocacy', 'protect', 'illegal fishing', 'hook and line', and 'gunman'.

He can no longer patrol because the beach owner has erected a fence along the shoreline. For now, his advocacy is limited to talking. There is much to be done. By the time he was shot, he says, the compressors and illegal fishers were mostly gone. Now they are back. Every night he can see compressors and purse seiners and trawlers fishing where they are not allowed to. The Philippines has banned bottom trawling in municipal waters, an area equivalent to the country's land mass.[35] And illegal fishing still happens. There is always more to fight.

A short film about Norlan by a Cebu TV station described him as 'the fearless steward'. Fearless stewards have families so I ask him if his children support him. 'Yes. They understand that they got their education [school fees] because of the sea.' Even after he was shot, they did not ask him to stop. 'Maybe because I had warned them that something might happen to me. So they did not oppose me. Because of what happened to me many fishermen in my village were emboldened to protect the sea.'[36] At the close of my interview with Norlan, I thank him, and he thanks me back. 'I want to share my story not just in the Philippines, but in other places too.' Eyes on the sea, as wide as can be.

© Christopher Hilton

The Aldeburgh scallop

12

A Very Slavery Job Actually

States should ensure that fishing facilities and equipment as well as all fisheries activities allow for safe, healthy and fair working and living conditions.

UN FOOD AND AGRICULTURE ORGANIZATION,
CODE OF CONDUCT FOR RESPONSIBLE FISHERIES[1]

The boat was older than their children and nearly as old as they were. They saw the rust on it, the bodged repairs. They were uncertain and worried by the state of it, but they had no choice: they had a contract and unless they went fishing, they would not get paid. They all needed the money because it was more than they would get at home, where they could find no work. The skipper and first mate were both Scots and both unfriendly. The Scots had supplies: coke and bottled water for them, but nothing for the crew.

Instead, the new crew, freshly arrived on this old scallop dredger, had to find what water they could. Then they were set to work, hard. The days were so long that they erased the nights. Work, work, and more work: dropping the scallop jaws overboard where they would scrape the seabed, squashing starfish as if they were plants, smashing animals, hauling, sorting, dropping, hauling, sorting. There were no breaks and hardly any rest. The skipper communicated by shouting. The men, who at first had only their shared situation in common, were yelled into their place, and their place was at the bottom of the heap on the 40-year-old boat that was held together by shaky welds and bad faith.

The boat was the *Olivia Jean*, a trawler built in 1980 that flew the flag of the United Kingdom. She was nearly 35 metres long and 7.2 metres wide. From the exterior, she looked old and tired and like she had not seen much love and care during her 40 years of service. Certainly love and care has not been a feature of the *Olivia Jean* or life on board it. Instead it was, for many of the men employed on it, a kind of hell.

There has been much written about modern slavery in fishing in recent years. Mostly, this involves reporters travelling far away to Asian or other fishing fleets and reporting on hideous conditions, running serious personal risks to find slaves or former slaves. Such work is impressive and worth every prize. All I had to do to find three former fishing slaves was to go to the next suburb.

I first heard of the *Olivia Jean* a couple of years ago, when a seafarer welfare contact told me that there were ex-crew in a safe house in Leeds. A safe house for seafarers? In my landlocked city? This seemed an extraordinary thing. I didn't get to meet those crew and cursed myself for a lost opportunity. Never mind, because a year or so later, three more men arrived who had fled the same boat and sought refuge in the improbable location of Headingley, Leeds, a neighbourhood known for its cricket ground and university students, not its escaped migrant fishing crews.

I arranged to meet the men through Stella Maris, a Catholic maritime welfare organization. I knew of it through writing about the shipping industry because it looks out for the welfare of commercial seafarers. The UK government does not record numbers of migrant fishermen, but they are thought to make up half of UK fishing crews.[2] Even if your fish is caught by UK vessels, which is statistically unlikely, it is not Captain Birdseye who is providing the fish in your fish and chips, but Isaac or Aditya or Pedro from Ghana or Indonesia or the Philippines. There are so many migrant fishers, and they are sometimes treated so carelessly, Stella Maris has had to widen its reach to include them. A meeting was arranged. I was early and kept an eye out: I was looking not for fishermen but for Black men. There aren't many in Headingley,

A Very Slavery Job Actually

and a trio of African men (the fourth had already left Leeds) would be distinctive.

They came. Nicholas, Peter, and Joseph. They are all family men, they all come from Ghana, they all wish they had never heard of the *Olivia Jean* or TN Trawlers, the company based in Annan, Scotland, that owns it. Sometimes when fishermen's welfare organizations or unions say the name 'TN Trawlers', it sounds like a curse. Try to invent the shabbiest, most fly-by-night fishing operator, and you would come up with something that looked like TN Trawlers. TN stood for Thomas Nicholson, a lorry driver from Annan who looked at the fishing industry and thought he could do that, and so he did. Years ago, a man in the Fishermen's Mission pointed to the town of Annan on a map and said to me, 'There's a terrible company based there.' He meant TN Trawlers.

In Ghana, where they had applied for fishing jobs through an agency, the three fishermen knew nothing of this. How could they know from over there whether a company had a good or bad reputation? 'It's difficult to get a job,' says Joseph. 'I have to rely on the agent to tell me, "This is what I have for you."' Fishing is a common occupation for Ghanaians and they are proud of their nautical college. 'You go there for a short course, they will train you something about safety and survival, firefighting. Then you take it from there.' The fishing at home has diminished due to competition from industrial foreign trawlers. And the pay on even a rusty scalloper like the *Olivia Jean* seemed adequate: £1,200 ($1,490) a month. Joseph had been working on a cargo ship as an AB, an able-bodied seaman. The job was OK: he travelled around the world, the pay was fine, by the end of it he was still able-bodied and a better seaman. Then the company went bankrupt and the bank seized the ship. Nicholas was an engineer on oil rigs. That was also a good job and he liked it. He got excellent safety training: 'I know that the UK, when it comes to safety, they are top. When anyone comes to train us, they are always from the UK.' On the rigs, there were standards: 'If you need to lift anything heavy, you don't do it alone, you ask for help.' Also, he had never heard of anyone paying an agency just to be employed. The cheek of it.

So he was surprised that he had to pay the agency about £250 ($310) to get a fishing job. Such payments are illegal under the International Labour Organization's Work in Fishing Convention,[3] which the UK has ratified but Ghana has not. Anyway, the men would not be UK employees, although they would be working on a UK fishing vessel. Instead, they would get transit visas and be of uncertain immigration status. Transit visas are designed for transit, usually for seafarers from abroad who need to board ships in the UK that then leave its waters. In a recent briefing paper on transit visas, the International Transport Workers' Federation (ITF) called them 'the starting point for the labour abuse of migrant workers working in the UK fishing industry'.[4] In 2018, it stated, 'every case of modern slavery in the UK fishing industry related to victims who had arrived on transit visas'. This is because a transit visa leaves them peculiarly vulnerable. 'As soon as they have "left the UK" (been on one fishing trip outside the 12-mile territorial water limit) they are no longer considered in transit. When they return to port, they are not technically in the UK because they have not come ashore. This situation is exploited by owners to pressure migrant crew into working days off, when in port and when they are supposed to be resting. These things all increase the risk of accidents at work.' They are not allowed to fish in the UK's 12-mile limit, and they are not allowed to work when they come into port. In practice, it means skippers refusing to allow their transit visa crew to disembark at all. It means the fishing boat becomes a prison.

The Ghanaian fishermen understood that the agency fee would be taken from their salary after the first month. Nicholas wasn't happy about this. But he needed a job and he had faith in the UK, a country that does not allow the upfront payment of agency fees, but Nicholas didn't know that. Peter thought the pay was bad, 'but half of something is better than nothing'. They all paid, they all flew to Heathrow, two arriving on 2 October 2020, the other a day later. They all had the same job and the same job description. Deck TN. Fishing trawler. Scallop boat.

There was no welcome. They were given instructions to go to a hotel and wait. On the second day, a taxi arrived and took them to Shoreham, where the *Olivia Jean* was moored. As Peter remembers it, 'We put our luggage down and then we went to work.' They worked throughout the day, throughout the night. 'We slept for two hours, we wake up, we work.' They had to prepare the scallop dredges because the *Olivia Jean* was going back out. 'When it was 12 o'clock we had to stop for a while and eat because we started work at 8 o'clock in the evening.'

Once the vessel left port, their 24-hour rhythm became clear. Their life would be work. 'You work throughout the day, and you sleep about two hours, and then you have to come back. In those two hours, you have to cook, and if you want your shower you have to take it then.' Nicholas corrects Peter. No, it was four hours. You work for 16 hours and get four hours, and the day is 24 hours. Whatever. Either way, you cannot get a chance to sleep. 'In those four hours you have to do other things. You have to wash your clothes, prepare your food.' They had been told that the skipper's mate – this is an actual job title, meaning second-in-command – would cook for them. Untrue. More of the four hours gone then, for food preparation and fuelling.

They are talking intently, and I am trying to compute this level of exploitation.

'So you don't get eating breaks in 16 hours?'

'Of course not,' says one. Another clarifies. 'In the 16 hours you work, you can't even relieve yourself. By the time you are going to find a place, the skipper is pulling in the net and once he is doing that, you don't have time. So say, once you've finished organizing the scallops, sometimes I can hide myself at the back. But in oilskins, it's difficult.'

If *Olivia Jean* had been a car, she would have been scrapped. She was rusty. Things were patched up. On a vessel at sea, patching up and cobbling repairs on heavy machinery can be done right or recklessly. On a scallop dredger, there are endless cables and heavy metal objects that can kill a human. On a boat so old, things were always going wrong.

'We shoot [the chain bags attached to the dredges] for four hours,' says Nicholas. 'Then we bring it up.' But the vessel was so clunky it would always bring up problems along with the catch. Chains that slip, links that break. 'Then you have to fix it before you shoot it.' Also if there are new crew aboard, you have to teach them too. None of that can be done in working time, so your four hours 'off' are really four hours for repairs and teaching new crew how to do repairs and how to operate. 'There is no money for that teaching. And also there was shouting, at times "fucking" and "fucking asshole".' Nicholas sounds shocked at this. All the men are Christians: a church committee is partly what is keeping them sane in their bureaucratic limbo, currently being endured in a coffee shop near a shopping centre. And 'church' is also Bryony and Steve from Stella Maris, who are sitting here too, because they are caring for the men and make sure to come from Hull once a month to take them out, to arrange something, to put a spike in the long, long days of tedium: with their current status, they are not permitted to work.

At sea the men tolerated all of it. The abuse, the racism, the discrimination, the working hours that looked nothing like official regulations. If they calculated their salary against the hours they worked, they would be earning £2 ($2.50) an hour. They tolerated it when the skipper and mate bought bottled water for themselves and some Scottish crew who were there for the first week. Bottled water was essential; you couldn't drink the water from the tank. As soon as the Scottish crew left, 'they stopped buying that water. They only bought it for themselves.' Nicholas shows me pictures of the water they were supposed to drink. It is brown. Nicholas couldn't stomach it, but they still cooked with it. They complained to the skipper and a drum of water was brought, then out on the fishing grounds the boat rolled and the water spilled. The men refilled it each time they were in port, but they still had to use the ugly water to cook. 'I saw all this happening,' said Joseph, 'and we knew our contract was 12 months. If I stay on this vessel for 12 months, I don't know what kind of a human being I will be.'

A Very Slavery Job Actually

They thought the price was high but worth it, until it wasn't. 'Calling us all sorts of names and other things,' says Peter, 'but it is better to stay alive than to die.' This is the second time he has used this phrase. I wonder if they were scared as well as shocked and insulted, that this could be happening on a British vessel, with a Scottish skipper, a few miles from the shoreline of the British Isles, where they could buy ice creams and plastic buckets if they were allowed ashore or any time off. 'Of course, I was scared,' says Peter. And then things got worse. One of the tow bars holding the scallop bags detached. 'We fought to get it back, but it was not available.'

Peter says this quietly, but imagine the noise behind that quiet sentence. Curses from the wheelhouse. The crash of waves and wind. Cries from the crew watching the tow bar disappearing. And then, relief. The boat could not operate without both tow bars. Maybe they could sneak ashore while a replacement was found. Maybe some rest, some good food, and a respite. But in port a truck was waiting with a new tow bar. That is, a new second-hand tow bar. It was 7 p.m. and the skipper told them they had to work all night to fix it because they had to go to fish. They worked, they tried, but they knew it wasn't safe. Their explanation isn't clear, but finally I understand that the tow bar should be secured with wire, but the skipper and mate instead made them fix it with rope. The men knew this was unsafe.

They had been sure that they would go to shore and get it properly mended. 'But when we got back,' says Peter, 'they just wanted to go out again. We said, no, we are not going with you. That is what triggered us.' 'It's a very slavery job actually,' says Joseph. 'I'm sorry to use that word and it's something when you sit down and recognize all these things, it's something painful.' 'We said we are not going,' adds Nicholas. 'Because we thought that the tow bar would kill us.' In 2019, an Indonesian crewman was killed on the *Olivia Jean* when he was hit by the tow bar.

The men's belief they would be killed was as strong as their belief that a job on a UK fishing boat would be safe and that they would be respected. The tow bar would kill them. It would swing round and slice a head open, or heft someone overboard. No doubt about it. They understood the skipper's urgency to keep fishing. He and the mate were paid in shares. The more scallops, the bigger the share. The Ghanaian crew were on wages and they would not be paid until the following month. 'All the skipper is thinking about,' says Nicholas, 'is more scallops. More scallops.'

In a report to the US Congress by the Task Force on Human Trafficking in Fishing in International Waters, the authors spoke of 'an especially vexing challenge'.[5] The challenge is to stop fishing vessel owners exploiting people from other countries to make a better profit. It doesn't seem vexing to me. It requires a regulatory environment that does not allow migrants to be employed on unsound contracts or transit visas that imprison them on board fishing vessels for months or years. Employment agencies that don't charge upfront fees. Skippers who don't run boats where crew get injured or killed.

A devastating Associated Press (AP) investigation into labour exploitation in the Hawaiian longline fleet, the same vessels that catch tuna and fancy fish for fancy restaurants in Honolulu, found that hundreds of Indonesians and Filipinos and others in the fleet were unable to go ashore because they had no visas. Instead, they had some spurious temporary permit that Hawaiian Senator Daniel Inouye had allowed the fishing industry to use for its workers, while also somehow controverting the US law that requires 75 per cent of crews on fishing boats to be US citizens.[6] Transit visa, temporary permit, same loophole, same leeway.

It is too easy to find reports and investigations of abuse, slavery, trafficking, because there is so much of it and because we are not yet rid of it. In the excellent book *Fishers and Plunderers*, these are some of the subheadings in the 'Abuses and Slavery at Sea' chapter: Abduction. Abuse, General. Beatings. Children. Death. Exploitation. Imprisonment. Murder.[7] A report by the Environmental Justice Foundation included a handy map of the cases it had covered and where: Cambodia (workers exploited by brokers); Hawaii (forced labour on long liners); Thailand (trafficking convictions); United Arab Emirates (bonded labour in artisanal fisheries); Sierra Leone (observers prevented from working); Myanmar (workers tricked into the fishing industry); Vietnam (uncontrolled fishing fleet); Spain (crew stranded on boat); and the United Kingdom of Great Britain and Northern Ireland (trafficked fishermen).[8]

In late 2022, members of the Liberian Coast Guard, stationed on the Sea Shepherd vessel *Age of Union*, boarded a trawler flagged to China named the *Guo Ji 289*. The Coast Guard found what a Sea Shepherd staffer called 'a little ship of horrors'. The Liberian crew had to sleep on filthy mattresses or cardboard boxes while the Chinese officers had decent cabins. The Coast Guard also found four cats on the trawler. This is not unusual: cats are kept to keep rats down. 'It was made clear to me,' said Sea Shepherd's Peter Hammarstedt, 'that these four cats were destined for human consumption in order to add variety to a daily meal of eating fish discards, the marine life that the fishing master deemed unsuitable for sale. They ate cats for lunch.' The website address for the account of this boarding ends with 'latest-news/they-ate-cats'.[9]

The *Guo Ji 289* was transmitting its AIS as was mandatory, but its AIS identified it as the *Zhe Dond Yu 83032*. There is no such vessel except in the input data of the *Guo Ji 289*'s AIS device ('AIS spoofing' is easy if you know how to falsify the data your AIS device is transmitting). Technology is currently considered a powerful solution for all sorts of malpractice at sea and it is certainly impressive. Global Fishing Watch, for one, provides data for free for anyone. I can track the *Guo Ji 289* and its last six months of fishing effort and port calls for no cost. Without satellite data, the *Guo Ji 289* could not have been traced because it was camouflaging itself with technology. Sea Shepherd and the Liberian Coast Guard, or any coastguard, do not have the money or resources to board every vessel in the massive Chinese distant-water fishing fleet. The US Department of Labor has now added fish to its list of goods produced by child labour or forced labour in China or by Chinese people outside the country. It has not added 'United States' or 'Hawaii' as places where fishes are caught by forced labour.[10]

The journalist Ian Urbina, who founded the Outlaw Ocean project, spends plenty of time now boarding awful-looking fishing vessels and finding that their crews live in awful conditions. In Outlaw Ocean's recent five-year investigation into the Chinese distant-water fleet, it found that in the port of Montevideo, Chinese vessels in the greedy and growing squid fishery 'have been dropping off at least one dead body every two months'.[11] The AP investigation, a year in the making, found Burmese slaves imprisoned on an Indonesian tropical island, as well as enslaved fish workers who had been sent to work on vessels that then sent their catch to Thailand. Thailand is the world's biggest processor of pet food, so I am probably feeding slave labour to my cat every morning. The AP writes of 'tainted seafood', and if only the taint were visible, the seafood would not end up on supermarket shelves, as it does. 'The slaves interviewed by the AP had no idea where the fish they caught was headed. They knew only that it was so valuable, they were not allowed to eat it.'[12]

Steve Trent is director of the EJF, which regularly produces reports into how poorly humans and fishes are treated in the fishing industry. The EJF broke the story of widespread abuse on Thai fishing vessels in 2015, with a report called 'Thailand's Seafood Slaves'.[13] 'That was the most egregious, there was nowhere else in the world where you had the regular and repeated use of extreme violence including murder.' He is using the past tense because he thinks the Thai fleet has improved. It's not perfect but it is better. For Trent, now China is the '800-pound gorilla in the room'. Humans are the worst marine predators, and he thinks China is the worst of us. Trent and the EJF are planning to target 'choke points' to eradicate slavery at sea: the unethical labour agencies and brokers that employ Filipinos and Indonesians and the three Ghanaians in Leeds. China has no choke points. Who can stop it? 'My analysis is pretty simple,' says Trent. 'I think we need to make this ping on the radar in Beijing in a way whereby they get to the point where they think this just is not worth it.'

I find an arresting calculation in a paper called 'Obliged slavery and the race to fish'.[14] The United States has a national slavery presence of 1.8 victims per 10,000 people. Using that, the authors worked out that 'seafood imported into the US has an average potential slavery risk of 3.1 kg per ton, 17 times higher than the risk of seafood sourced from domestic fisheries'. Being sure that your fish supper has not been caught using slaves is, write the authors, 'practically impossible', and that impossibility starts at sea. Trans-shipment makes it easy to mix seafood, to fake fish. It is impossible to oversee.

The Marine Accident Investigation Branch's latest annual report showed that 31 people had been injured or killed on UK-registered fishing vessels.[15] Those are only the accidents that are reported. The ITF encounters many cases where injured or dead crew members are sent home without any report being made to the MAIB, or where the report is made when the migrant is long gone. 'The situation back home must be absolutely incredible before they come and work in our country for what they're getting and how they're treated,' says Doug Duncan, a chaplain with Stella Maris in Scotland. 'For me, it's a disgrace.' Good people in fishing use this word frequently when it comes to TN Trawlers. Alastair Robertson, a skipper and boat owner in Scotland, told the BBC that 'it's a disgrace, anybody that abuses their crew'. He employs migrant workers, and they are paid and treated the same as anyone else. That is how it should be.

Labour can make up two-thirds of the operating costs of a fishing vessel and foreigners are cheap. A *Financial Times* journalist reported a social media conversation between the skipper of an Irish boat who had posted seeking crew. '*Strathmore* looking [for] a man for the prawns for a trip,' the skipper wrote. 'Men are like hen's teeth,' replied a friend. 'Tell me about it,' the skipper replied. Foreign crews will go to sea when young British and European men will not.

Nicholas is talking again about his life on the rigs. He cannot help but compare and it sounds like a kind of mourning. 'Where I worked, you go for 28 days on, 28 days off. If I have to go for a course, they pay me for that. But this, you just work all day. There [on the rig] we have a gym, we have table tennis, we have a medic on board, we have everything you need on land, so we feel at home. But here there is nothing. Empty. You have to give us some time before we go fishing. One time I needed a drink. He said we were going fishing so I had less time, so I had to quickly jog [to the shop] to get the drink. I thought, what kind of life is this? My contract states that they have to give us sufficient rest time but we are not getting it.'

The nature of fishing makes the job an isolating one. To this, add the behaviour of the captain and first mate, and the business practices of the owner. All can make life even worse. 'On *Olivia Jean*,' says Joseph, 'there was Wi-Fi, but the skipper switched it off whenever he liked.' 'Especially if we don't get enough scallops,' says Nicholas. 'Sometimes we have to phone our family but the Wi-Fi is off. When you are looking for the Wi-Fi signal you can see that their Wi-Fi is still on.'

In Shoreham on 4 November 2020, the Ghanaians' immigration status must have been on their minds. It would be on their minds for the next three years and counting. They were in port, so they had a phone signal, and one of them had a number for Paul, a chaplain working with Stella Maris. Paul was too far away, so he contacted another chaplain in nearby Southampton. Nicholas: 'We were taking on fresh water and we were about to start sailing again. Then the police came.'

A local news website reported the events. 'Five "slaves" rescued from fishing boat in Shoreham Port'.[16] The 'rescue' was imperfect. The men weren't allowed to collect their belongings, not even their phones. This upset Nicholas because he knew he had a copy of his contract, which would be relevant, in his bag on board. He had read it that morning. When his bag was returned, there was no contract in it. Another newspaper was impressively localist in its reporting. 'It is understood that neither the people on board the boat or those arrested are from Sussex.'[17]

The people who were not from Sussex were separated according to guilt. Two were arrested on suspicion of modern slavery. The other five were also formally interviewed at the police station then taken to a hotel. The men's account now gets confused. The luggage was brought after two days or was it four or five? The police took them to the hotel but then they moved with Stella Maris to a house and the police didn't know where they were. I understand this confusion. It was a stressful time. All that matters is that they were going from one bad situation where they were under threat to one where the threat was different yet also stubbornly stressful. Sussex Police confirmed to me that they interviewed the men once at the station and never again. Despite having five compelling witness accounts of poor treatment probably amounting to slavery, the police released the skipper and mate after a day and they were never prosecuted.

On 19 November, the three Ghanaians arrived in Leeds. Their Indian crewmate had a friend in Wales and went to stay there. Another crew member went home to Sri Lanka. The Ghana Three live in a shared safe house provided by a small charity. And Stella Maris began the arduous process of getting them 'conclusive grounds' to stay in the country. 'Conclusive grounds', if you talk to anyone who has been human-trafficked or worked as a slave, is the golden prize. It means you are believed, that you are free to go home or to stay, and that if you stay you can work. For now, the men are living with 'reasonable grounds', which were given when they were rescued. This means they are not deported, but they cannot earn money. They can exist but not really live.

I ask the men, in that bland coffee shop, whether they want to stay here. 'We want to work. We have families who depend on us.' They have a small weekly income of £65 ($80) from the Salvation Army, which gets it from the government for people in safe houses. That is not a standard amount. 'I know a fisher in Bristol,' says Bryony from Stella Maris. 'He gets £41 ($50). It's all different. We also had two guys from the same company in Scotland who were getting less as well. It's baffling.'

They cannot work, so what do they do? They volunteer in the Mind charity shop. They do voluntary work at a church, cleaning. 'Sometimes,' says Nicholas, 'you'll just be in the house, struggling.' The first time we meet, the men have been waiting for one year and six months for the Home Office to make a decision about their case. The normal processing time to reach a decision is six months but, as soon becomes rote, Covid is given as the excuse for the delay. Other men who were rescued from a boat in Scotland were taken off months after the Ghanaians and got their decision in five months. A previous crew which was also removed from the *Olivia Jean* only waited eight months.

I like Bryony's expression of frustration. 'I don't know what's keeping them.' Like the Home Office is delayed by traffic, not by either spectacular inefficiency or callousness about the hidden labour force that keeps the UK fishing industry going, on the cheap and on the sly. The money is not a problem for the men themselves, but for their families in Ghana. I ask how their families are and there is a chorus of exactly the same phrase.

'They are not OK.'

'They are not OK.'

'They are not OK.'

There is no Wi-Fi in the safe house. If they can afford credit, they can call. Nicholas's children are six and three years old. He hasn't seen the younger since he was one year old. He is now becoming very distressed. 'If I have capital, I will never travel again in my life for someone to treat me like this. I prefer to stay in my country, just to have something small, to get my peace. I will have my business, farming, so farming is a long-term investment, so I want to invest in animal farming. Additionally I will buy and sell electronics. If I get the opportunity to do that, I will stay in one place and live with my family and live a peaceful life.'

Everyone I speak to about TN Trawlers is furious about their callousness and bewildered by their relative impunity. Bryony, who has been a port chaplain for years and therefore in my eyes is a national treasure, said firmly, 'That company is wrong, it's criminal on every level and everybody's been aware of this since 2012, and even before that

A Very Slavery Job Actually

we knew about the deficiencies on the boats and all that kind of stuff, and yet, everyone in the industry is aware of it as well. At a conference last week, a guy said, I know sooner or later the words TN are going to come up. Everybody in the industry knows them, all our charities know them. How are they still working?'

I compile a database of TN Trawlers incidents. Partly because there are so many, I get them mixed up. There are only so many 'crewman injured' accounts I can keep straight in my head. It is probably not comprehensive, but I find 23 up to 2020. This is just a sample:

2000: Crew serious injury on *Philomena*: crew member fingers crushed when he grasped a running wire that dragged his fingers into a block.

2001: Crew fatality on the *Philomena* when crew member was hit on the head by a tow bar.

2005: Crew serious injury: crew member required stitches to his forehead after being hit by a steel holding hook.

2008: Tom Nicholson fined £500,000 ($620,000) at Newcastle Crown Court after admitting 26 offences breaking quota regulations. Fish landed on 53 trips on two vessels valued at £260,000 ($323,000) but probably worth £1 million ($1.24 million).

2010: The Marine Accident Investigation Branch recommends that *Olivia Jean* stop fishing.

2012: Joel Quince knocked unconscious on *Philomena*, bleeding from head wound. Tom Nicholson told his son to carry on fishing for 12 hours. Even when they landed, Quince had to phone a seaman's mission to arrange transport to a hospital.

2017: Nine crewmen taken from *Sea Lady* in Portsmouth: eight Africans and one Sri Lankan. Tom Nicholson junior arrested. Boat moored after crewman suffered head injury.

2019: Crew member fatality. Indonesian crew member on *Olivia Jean* struck on the head by a tow bar when the vessel was 39 miles north-east of Aberdeen. He was airlifted to hospital but died 12 days later.

2020: Crew serious injury. Crewman on *Olivia Jean* hit by tow bar. Taken to hospital with chest injuries.

2020: Five men rescued from *Olivia Jean* at Shoreham.

2020: Crewman's foot crushed by tow bar. Crewman kept on board for 16 hours before landing ashore. Skipper gave him £20 ($25) to go to hospital. Foot nearly amputated.

Over the course of one month in 2012, six people jumped overboard into Plymouth harbour to escape TN boats.[18] Ten years after Joel Quince's head was sliced open, Tom Nicholson was prosecuted in court for failing to help Joel in time. Text messages between son (who was skippering the boat) and father (ashore) were reproduced.

Tom Nicholson junior: 'Is there any port I could drop him off at or just Troon? He's saying he feels dizzy and his eyes are going yellow but they do that all the time with these Filipinos . . . I don't want something to happen to him like but don't really want to stop fishing, is it too much hassle to get a coastguard out?'

Tom Nicholson senior: 'No do not call them. He'll be fine if bleeding stopped, give him hot drink and pain killers. Tell him you'll check it in morning.'

When his son messaged him the next morning to say he was bringing Joel ashore, his father responded with 'whatever'.[19] Tom Nicholson was fined £13,500 ($16,740) and had to pay £3,000 ($3,720) in compensation. For a company the size of TN Trawlers, this was a minor punishment.

'Everyone [in the fishing industry] wants to distance themselves [from TN Trawlers] now,' a fishing labour expert tells me. He is not allowed to speak publicly. 'TN Trawlers have featured in fishing associations, they were a big fishing family. Now they serve as a convenient guinea pig.'

A Very Slavery Job Actually

As I write this, TN Trawlers has sold all its vessels and attempted to dissolve the company, but this attempt was blocked because there were investigations and court cases under way in England and Scotland. (They have since been dropped.) I look at my database and think: 20 years and no one stopped this company from going fishing.

At a later meeting in a different coffee shop, the men look exhausted. Nicholas's eyes are red. He refuses coffee because he is not sleeping, and they tell me that they are all on mirtazapine, an antidepressant. They still cannot work; they are now living on £71 ($88) a week from the government, though the ITF helped them with a payment for their children's school fees. To get their money, they have a prepaid card that is free to use only in NatWest ATMs, and the nearest is two miles away, so they walk a lot. The church is a mixed blessing. They came to it when the son of the pastor heard them speaking and recognized their dialect. The men welcome the community spirit and it's nice to talk in their first language, but they cannot freely speak there. Joseph still has not told his family that the first three appeals were rejected. He has not told them that he cannot work. 'How can I do that? I am nearly dying here and if I tell my wife that, she doesn't work, she will nearly die too.' Church gossip is perilous for anyone keeping a secret.

They have a new lawyer, and I am cautiously hopeful without much reason to be. When I asked Sussex Police why the skipper and mate, arrested on suspicion of human trafficking, were released, the response was that 'there was insufficient evidence to bring about a prosecution', despite five witness statements and five willing witnesses. The Home Office has suggested the Ghanaians go home but, unlike their Sri Lankan and Indian crewmates, they will not. 'We have been treated unfairly,' says Nicholas, 'so we said no.' Despite being the subject of at least two long-running criminal investigations, Nicholson and company have never been found guilty of human trafficking or modern slavery. When Police Scotland raided TN Trawlers' premises, they forgot to check whether the boats and crew would be there (they were out fishing). Later, 60 officers found no evidence of human trafficking. 'The Home Office said later,' said reporter Chris Clements in the BBC documentary 'Slavery at Sea', '[that] two Indians, and a Ghanaian were found aboard the vessels – fishing in UK waters, in breach of their visa conditions. They were arrested, then given instructions to leave the country within seven days.'[20]

Joseph speaks now. 'We can't go home, the stigma will be huge. In Ghana, you have to understand, when people come home from abroad everyone asks them, "Where is my gift?" If I go home and I can't feed my children . . .' He doesn't finish that sentence. 'The Home Office say they are fighting modern slavery, but they are deceiving everybody. They have a clear example of modern slavery on the table before them, but the Home Office don't want to listen to us.'

The boat was awful, says Joseph, 'but what the Home Office is doing has made the situation worse'. The procedures grind on and meanwhile three men must spin around in their minds, because they trusted that they would be protected by United Kingdom law and they were not. 'I am dying slowly,' says Nicholas. The last time we meet, his eyes are slightly wild and I am worried about him. 'I am dying. I know within me, I know my strength and when I'm losing strength I know. I'm losing my mind.'

© Rose George

Mobile fishmongering

13

A Nice Bit of Salmon

*Animals shall be treated well and be protected
from danger of unnecessary stress and strains.*

ANIMAL WELFARE ACT, NORWAY

I did not come to Norway for its aquaculture. But in Norway it is unavoidable. Farmed salmon is Norway's third biggest export after oil and gas. There are 1,758 farms along its 13,000-mile coast and in its fjords.[1] There are also 20 aquaculture visitor centres, a startling number for someone from the UK, where there are precisely none, despite a flourishing Scottish salmon farm industry. (Plans to open one in 2018 on Skye sank without trace, although the aquaculture company Mowi allows tour groups to visit if they fill in forms correctly.)

So it isn't until a couple of days before I reach Bergen, and after 10 days in the country, that I think of organizing something. The Norway Fisheries Museum offers school trips to visit fish farms, including a trip on a fast rigid inflatable boat (RIB), but though I'm welcome to join the group later for a talk and a bit of fish dissection, the RIB is full. I write to Lerøy, a huge seafood company that sponsors the Storeblå aquaculture centre in Bergen. (Many aquaculture visitor centres are sponsored by the salmon industry.) I expect my email to sink without a bubble, then someone called Øyvind responds and says he can take me out to Buarøy salmon farm and is a meeting at 8.30 a.m. on a faraway headland acceptable? The meeting point is a shop that backs on to a small marina where there are sensational views of snowy islets in the fjord. There is no Øyvind and no answer when I call. I stand outside the shop, watching for a car, until Øyvind calls back and says, 'I'm behind you,' and I finally realize he has come by boat. Of course.

Øyvind is from a fish family. His grandfather and father were fishermen, along with his brothers. Yet first he trained to be a bricklayer. I look at him with puzzlement. Why would you do that? He makes a shrug expression. I don't know. When you're young, you are stupid. A few years of bricklaying, and he came back to sea. He is now the operations manager for several fish farms around Bergen. Buarøy is aquaculture site 3108 on BarentsWatch, the Norwegian government's information portal on coastal activities. It takes us 10 minutes on a comfortable warm boat to get there. I've never seen a fish farm up close before. On the way to the island of Hitra, I drove past a site but I was concentrating too hard on driving in driving snow and all I noticed was that the rings looked ugly in the grey water. Here, as the boat slowly steams up to the ugly rings, I don't think that. I think: the fishes are jumping! 'They have to do that,' says Øyvind when I express pleasure at the leaping. 'It's how they get air.' Why else would football use the expression 'he leapt like a salmon'?

Øyvind gives me facts as we slowly pass the salmon pens. The pens go deep: their 'skirts' (the walls of netting) can descend 40 metres. Divers are employed to fix holes in the net as long as they are not too deep, but otherwise monitoring is done by a fixed camera and ROVs. Holes and frays happen often enough, either because of rotting caused by lice treatments, thinks Øyvind, or because bluefin tuna, now plentiful around here again, 'go through it like a torpedo'.[2] The world record for running the mile is 3 minutes, 43.13 seconds, which works out at half the speed of *Thunnus thynnus*.

There are also local sharks called pig horns, less than a metre long and still trouble for fish farmers. The sharks nibble through the net rather than bulldozing but the result is the same: escaped fishes. There are 180,000 salmon in the pen I'm watching. Not all the dozen pens are in use; they are shut down every couple of years to let the seabed recover from the volume of salmon waste. Each pen holds different ages of fish. It takes

A Nice Bit of Salmon

2 to 3 years to raise a salmon from roe (egg) to alevin (toddler) to fry (child), to smolt (teenager) to dinner. This is done by reproducing the wild salmon's anadromous lifestyle. Anadromous (Greek for 'up' and 'running') means salmon are both freshwater and saltwater: in the wild, salmon are born in rivers, head to the sea to live, then return to the rivers to spawn and usually die. Farmed salmon are kept in freshwater for 8 to 15 months and move to the pens when they weigh 500 grams.

Norway is where salmon farming began. There had been entrepreneurs hatching salmon before, in the United States, Canada, and Scotland, but the aim there was to repopulate rivers where the wild salmon population was diminishing because of human activity. Salmon has always been a popular fish to consume: it is nice and pink, it yields a robust fillet that doesn't fall apart in a frying pan, it's not too bony, and it has a mild and amenable taste. The BBC's Good Food site calls it 'the easiest of fish to cook'. When I once spent a day working as a mobile fishmonger, we had plenty of fishes for sale. Sea bass, hake, haddock, sea bream, shrimp, fishcakes. It was a cornucopia in the back of a Ford van. All day the most frequent request was for 'a nice bit of salmon'. Tony Walker, the fishmonger whom I was shadowing for the day, attributed this to the hot weather and had stocked accordingly. Even in rain, salmon is one of his most popular products. The world's desire for a nice bit of salmon is a problem that goes beyond weather and beyond Tony. Everyone may want salmon, but where will it come from?

Norway – population 5 million humans – produces half of the world's farmed salmon.[3] This long country has fjords and a huge coast. It has farmers whose farms are near water who can install a single fish pen to supplement their land-based efforts. Fish farming is not easy. It is not simple to mimic nature. And if we could, hardly any farmed fish would survive: many wild salmon don't survive the journey to the spawning grounds or spawning, and wild salmon have plenty of predators.

Slowly fish farming became a corporate affair, and now it is run by a handful of big companies, mostly Norwegian. Mowi is the biggest, then Lerøy, Bakkafrost (Faroe Islands), Cooke Aquaculture (Canada), then Chileans, who produce almost a third of the world's farmed salmon.[4] That prestigious Scottish salmon that you buy? It is farmed in aquaculture sites that are 100 per cent foreign-owned. The last independently owned Scottish aquaculture site, Wester Ross, was bought in 2022 by Mowi, a Bergen-based company with 13,000 employees and revenue of $5.16 billion.[5]

In the 1970s, when Norwegian fish farming began, aquaculture supplied 4 per cent of fishes eaten by humans.[6] Now it is more than 50 per cent, although China is the largest aquaculture producer and China routinely inflates its figures. The FAO aims to grow aquatic animal production by 35 per cent by 2030.[7] It also thinks world capture fisheries will recover. For FAO and plenty of other organizations interested in where the world's food will come from, the future is blue. Blue economy, blue revolution, blue farming, blue transformation. And everything blue is rosy.

Once I've seen several dozen salmon leaping, I've seen them all. So we head for the barge, which is the site's floating office. It looks like nothing much but inside it is comfortable. Yes, said a marine biologist I met later, Buarøy is one of their best. A show barge. The office is Scandi-style, with pleasant furniture, two appealing lounge chairs, and plentiful coffee. A bank of monitors on one wall enables staff to watch the pens, and for me is excellent TV. I am lured by the smooth swimming of the salmon. The footage is black and white, and this is more realistic because farmed salmon are grey.

Wild salmon get their pink colour from eating krill and prawns and other crustaceans that contain a carotenoid called astaxanthin. Farmed salmon get coloured by pigment or astaxanthin added to their feed. No one would buy a nice bit of grey salmon. Customers associate a redder colour with a better fish and will pay more for it, so the industry uses something called SalmoFan, which looks like a paint chart but for fish.[8] Academics have spent an impressive amount of time working out which colours sell best. Using local soccer teams and choirs as testers, a Norwegian study found that colours R23 to R29 were preferred. Salmon coloured at R22 or under 'were difficult to sell at any price'.[9] I'd try to explain what those numbers represent, but it would just be more or less 'salmon colour'.

Øyvind loves his job. But he is candid about the difficulties facing his industry, which he summarizes by repeating questions he is asked at social events.

'But what about the lice?'

'But what about the fish escapes?'

'But what about the environment?'

This last question is common. The volume of fish raised in aquaculture sites can overload the ocean with nutrients, just as human faeces can overload rivers. The Norwegian Climate and Pollution Agency calculated that waste from an average-sized fish farm equalled the sewage of a city of 50,000 people.[10] I can't see the sewage on fish TV, but other things are going on. Look, look, a lumpfish has stuck its suckers to the camera.

Aquaculture does 'co-culture' farming using lumpfish and wrasse. These fish work as cleaning staff, as both are willing to eat the sea lice that prey on salmon. The lumpfish has stuck to the camera because it likes a flat surface. Otherwise they have to swim all the time so they are not dragged into the walls of the pens, and sometimes they need a rest. This can be on a camera lens, but they also have an artificial habitat. That sounds splendid, as if there are giant waving ferns and trees and coral and seaweed. But the habitat is a banal black plastic thing. It doesn't look like much, but as long as the lumpfish can hide and rest, that's enough. Or that is enough for fish farmers.

If salmon farms were planes, Buarøy would be a private jet. It is clean, it is smooth, and it is very quiet. Only six people work here permanently, but when things need to be done, mobile teams are summoned, and things often need to be done. The delousing team. The slaughtering team. Today is the day for the fish-the-lumpfish team, says Øyvind, and points out of the window at men on a boat with big fishing nets. It is an industrial adult version of a child at a rock pool. Although cleaner fish sometimes make up 10 per cent of the pen populations, currently it's about half that. But 5 per cent of 180,000 is still 9,000 lumpfish. These men will be fishing for days. The lumpfish have to be removed because delousing is imminent and that will kill them, so they will be killed anyway. Overdosed, says Øyvind. He is sure it is a humane and gentle death and better than the fate of most wild-caught fish. But wild lumpfish live for up to 8 years. The oldest recorded lumpfish was 13.[11] No chance of that here: a year's work and they are done. They never see adulthood.

Me: It's not much fun being a fish in Norway, is it?
Øyvind: Not really.

Cleaner fish began to be used in salmon farming in the 1970s. It seemed a perfect solution. Salmon are prone to lice and diseases. If you were going to design a perfect fish to be farmed, it would not be a migratory lice-prone carnivore that requires feeding. Antibiotics and delousing chemicals are expensive. They linger in the water. They affect other marine life, and the lice were getting resistant. And here were pliable animals free for the taking who enjoyed eating sea lice. In the ocean, salmon would rarely be attacked: the ocean is big, the lice are floating around looking for a fish that makes up a tiny percentage of the ocean's inhabitants. But stocking 180,000 salmon in a pen is like building a sweet shop for lice. They are nasty parasites. They don't like scales, so they attack the salmon's head and neck, which have none. Salmon are routinely found with skinned heads.

Where do the cleaner fish come from? Both wrasse and lumpsuckers are hard to grow in a farm setting. So aquaculture's solution is problematic, as it requires fishing them from the wild. Off Scotland, wrasse are caught in creels because both wrasse and lumpfish like to hang out on the seabed. In fact, the behaviours of cleaner fish are nothing like the natural behaviour of salmon. They also have to be fed different food for the three weeks it takes them to start munching on lice.

A Nice Bit of Salmon

Probably more than 60 million cleaner fish are now used every year.[12] Cleaner fish are marketed as a more sustainable solution than antibiotics. But it's also a marketing solution: antibiotic-free salmon sell for 30 per cent more on international markets.[13] When the Institute of Marine Research in Norway surveyed 500 Norwegian fish farms, it found only 'a weak and variable effect of using cleaner fish in the battle against salmon lice'. Use them early on and they could delay the need for other delousing procedures. But generally, 'farms which deployed large numbers of cleaner fish achieved a clear effect for 10–20 weeks, but then louse numbers started to rise again. That suggests the cleaner fish either stop eating the lice, escape, or die within a relatively short period of time.'[14]

The industry does not hide its lice problems. If you ever want to encounter a giant louse, you should head to the Storeblå aquaculture visitor centre in Bergen, because its three-metre-tall model of a louse is a star attraction. An in-your-face louse: it must be like being a trapped salmon. At Buarøy, they count the lice every Monday by hand. They lift out 20 fishes, put them in tanks, and physically count their parasites. Most lice are visible to the human eye though they are not three metres tall. The lice hunters are looking for the following categories in order of priority: adult female louse with or without egg strings; moving (large and small); attached. I wonder whether an attached louse is better than an unattached one: cut loose, the lice and especially the larvae float on the sea currents and can infect other facilities or wild salmon populations.

There are many videos online of salmon horribly infested with sea lice, and the lice tails drift in the water like a woman's scarf from an open-top car. The Norwegian government requires salmon farms to keep lice levels to half a female louse per fish.[15] This is stricter in early summer, when the wild salmon population is migrating. Delousing can be done mechanically, thermally, with hydrogen peroxide, using freshwater, using medicine, or some combination of those. Thermolicing – dunking the salmon in a bath of water up to 34 degrees Celsius for a few seconds – works on the lice's dislike of temperature changes: the heat is meant to kill them. But salmon don't like the heat either, and whatever the treatment, plenty of them die throughout it. Mechanical treatment aims to dislodge lice by 'flushing, brushing or turbulence'.[16] Temperature or mechanical treatment is more favoured now than chemical treatments because the lice have developed resistance.

Any louse treatment affects the fish. Their growth is retarded, their gills get damaged, they take longer to recover. Salmon farm critics provide images of salmon half-eaten alive by lice, with gouges and weeping sores. That is one way to tell an escaped salmon; another is to look for damaged fins or a rounded tail, both signs of swimming in confined circles against walls of mesh.

I ask Øyvind what fish are to him. 'They are creatures indeed,' he says in his charming English. 'I like to see the results of our production. If the fish are stressed, there is a decrease in the quality of the fillet. We call it gaping: you see the fillet is cracking. To get efficient production you need people to be safe and the fish need to be happy too.' Also, they taste better.

In the Outer Hebrides, there is a loch. Images of it show a perfect Scottish scene: still waters, rising hills. The loch serves as a boundary between Harris and Lewis, two island nations that share a land mass but try to forget they do. It also houses two salmon farms belonging to Mowi. Not every coastline is suitable for farming salmon: the fishes require a certain temperature and a decent current. Scotland qualifies; so does Iceland. In 2024, Loch Seaforth became notorious when at Loch Seaforth and Noster, two aquaculture sites of the 40 seawater farms owned by Mowi in Scotland,[17] more than a million salmon died over one 18-month production cycle. The Scottish government publishes detailed salmon mortality statistics for anyone who cares to look. From these, I see that Seaforth also had a terrible week in November 2023: first 206,563 salmon died from amoebic gill disease (AGD), proliferative gill disease, and 'treatment loss'. Only four days later, 151,275 deaths. Cause of death: AGD again and 'bacterial'.

The ocean is full of viruses and bacteria, and salmon confined to ocean pens are perfect prey. But a million dead fish?

In Norway, 100 million farmed salmon and trout died in 2023.[18] In cattle farming, the mortality rate is about a tenth that of salmon farming, but higher for calves.[19] The Sustainable Food Trust charity believes it is misleading to compare fish mortality with that of other farm animals. This is because 'a typical salmon will lay about 7,000–8,000 eggs, whereas a sheep will typically produce two lambs'.

Some diseases that farmed salmon are prone to are AGD, infectious salmon anaemia, pancreas disease, and cardiomyopathy syndrome. The Scottish government defines cardiomyopathy syndrome as 'an economically important disease of adult Atlantic salmon in sea water'.[20] Do they get cardiomyopathy because they are stocked so densely in the pens that their hearts get stressed and they can be identified in the wild by their damaged bodies? The Scottish government blames a viral aetiology. Nobody suggests that there is a welfare issue. There are no fish vets: the usual response to disease is to cull (a word only used to describe the slaughter of animals). But the European Council's directive on the welfare of farmed animals (including fishes and amphibians) doesn't seem aware of this. 'Any animal which appears to be ill or injured must be cared for appropriately without delay and, where an animal does not respond to such care, veterinary advice must be obtained as soon as possible. Where necessary sick or injured animals shall be isolated in suitable accommodation with, where appropriate, dry comfortable bedding.'[21]

In 2021, Changing Markets Foundation sent an investigations team to 22 Scottish farms, getting underwater footage from six of them. They saw 'salmon swimming blindly around the cages with missing eyes, or those with large chunks of skin and flesh that had been eaten away. Many of the fish were covered in parasitic sea lice that were eating into their skin, some had seaweed growing in open wounds. There was also evidence of gill damage, fin damage, abrasions and lesions, infection, and mouth damage.'[22] Photographs released by the

Scottish government in 2018 after a freedom of information request show horribly diseased animals.[23]

Under Scotland's Animal Health and Welfare Act 2006, fishes are included for protection from unnecessary suffering (as 'a vertebrate other than man'). But welfare rules for farmed animals exclude fishes, reptiles, and amphibians, although farmed fish is the fastest-growing form of farming worldwide. A '10-year health framework' for farmed fish maintains that 'healthy farmed fish have better survival rates and require fewer interventions by farmers', but a framework is not law.

The only protection that fishes have in English and Scottish law comes from earlier acts that guarantee them 'protection from unnecessary suffering'. Every minute about 9,000 fish are slaughtered in UK aquaculture sites,[24] but there is still 'little to no regulation,' according to the Conservative Animal Welfare Foundation, 'on the treatment of these animals'.[25] A few things that are not covered in UK law: how farmed fish should be delivered, handled, stunned, or killed.

The Farm Animal Welfare Committee is a panel of veterinarians and animal welfare experts that advises the government. Its remit is to oversee the welfare of farm animals, and its published opinions sound like a prosecutor listing charges against humans: lameness in sheep, mutilations in piglets, bone fractures in laying hens, beak trimming of laying hens, mutilations of sheep. In 2014, it briefly considered the welfare of farmed fish and hasn't touched the subject since. Until then, if the well-being of farmed fish was considered at all, it was whether the water was acceptable.[26] Now that it is mostly accepted that fish feel pain, it may be time to consider whether the water is less crucial than workers ripping open the gills of salmon while the fishes were conscious, as was observed by the NGO Animal Equality at a farm run by Scottish Salmon Company. Or fishes being ineffectively stunned and then having their gills cut and taking seven minutes to die, which is so common as to be routine. The Aquaculture and Fisheries Act (Scotland) of 2013 does not mention welfare. Under the Welfare of Animals at the Time of Killing Regulations (England) 2015, fish should be spared 'any avoidable pain, distress or suffering during their killing and related operations'. But 'avoidable pain' is not defined, and the UK has no requirements that fishes be stunned before they are killed.[27]

In 2021, a government spokesperson admitted that there was no routine welfare inspection programme of farmed fish at slaughter in England and Wales, and a freedom of information request to the Scottish government revealed that there is no process for regular inspections either.[28] When Animal Equality wrote to departments it thought had responsibility for farmed fish, asking if they performed inspections, each department wrote back to say it thought someone else was dealing with it. Norway has an Aquaculture Act that requires fish to be stunned before slaughter although only half of farmed fish get this treatment. Its Animal Welfare Act though requires 'respect for animals', which is nice.

The slogan of the Norwegian Seafood Council is 'Together, we are winning the world for Norwegian seafood'.[29] Norway provides all that salmon,[30] but Norwegians don't eat it. 'It's too expensive,' says Øyvind. This does not surprise me after days now of standing in Norwegian supermarkets staring in shock at the price of broccoli. Mowi in its latest annual report writes of leading 'the blue revolution', of course, but also that its aim is 'to provide a growing population with delicious, healthy and nutritious food from the ocean'. Much of that growing population, which will be in developing countries, can't afford salmon. A graph in Mowi's 'Salmon Farming Industry Handbook' shows the relative price differences of other protein compared to salmon. Even beef, the next costliest protein, is nearly half the cost of salmon. 'Salmon,' the report admits almost in a whisper, 'has historically always been a rather expensive product on the shelves.'[31]

At least half of the cost of salmon farming is food. For decades, 90 per cent of that food was fish: wild fish and forage fish (a strange term meaning fishes that humans scorn to eat) that became fish food. Now, still, half of all wild-caught fishes are 'reduced' into feed. I'll unpack that half: it adds up to a stunning 490–1,100 billion fishes (and that's not including discards).[32]

The ratio used to measure the practice of feeding fish to fish is known with charming mundanity as the Fish In: Fish Out (FIFO) ratio. At its worst, the FIFO ratio was five: 5 kg of fish produced 1 kg of salmon. By 2017, the accepted FIFO ratio – for all sorts of farmed fish – had dropped from 5 to 0.28. But then a paper in 2024 by Oceana found that accepted FIFO numbers were probably highly inaccurate because they didn't count 'trash fish' (whole fish that is edible but undesirable) or trimmings or fish oil. Depending on the type of fish in and the type of fish out, the FIFO ratio could be 27 to 307 per cent higher.[33]

The other relevant abbreviation is the feed-to-conversion ratio (FCR). The salmon industry often promotes salmon as having the lowest FCR, needing 1.5 kg of feed for 1 kg of body weight. Chickens are similar. Cows need much more food to grow.[34] But that 1.5 kg still contains small pelagic fish or – worse – krill, a crustacean often sourced from Antarctica, where it is a critical food source for penguins and seals and other key species.[35]

There are efforts to find alternative food sources for farmed fish. Black soldier fly larvae are promising.[36] Experiments include lupin kernels being fed to giant tiger prawns[37] and blood meal being served to European bass.[38] Better, obviously, to farm fish that are not carnivores. In fact, the most farmed fish in the world is the herbivorous grass carp, followed by the omnivorous tilapia. There are modest hopes for the Bermuda or brassy chub, both algae eaters, both eaten widely already, just not in Western restaurants. Some people might think them too bony, too different from the alpha fish salmon and tuna. But human preferences can change or be changed. And they should be. I'm not suggesting the world goes vegan, but aquaculture would be a proper blue revolution if farmed fish did.

At a sustainable fisheries seminar in Senegal, I watch a presentation by a charming Madagascan minister. His job title is Minister for Fisheries and the Blue Economy, and his name is Paubert Tsimanaoraty Mahatante, which he translates for his audience. 'Someone who does no harm.' Then he says, 'I hope we don't do harm to fish.' His cheeriness is as endearing as his long list of aquaculture endeavours that are being done in Madagascar, where the future is blue and green. I'll use the French names because they are fun. *Crabiculture* (crabs). *Crevetticulture* (shrimps). *Langousticulture* (lobster). *Ostreiculture* (mussels). *Algoculture* (algae). *Spiriculture* (seaweed). *Tilapiculture* (tilapia). *Holothurlculture* (sea cucumbers: giant slugs that are prized in Asia).

By 2035, aquaculture will be worth nearly $400 billion.[39] Investors now zoom towards aquaculture at tuna-speed. Aquaculture will be the future, but it shouldn't necessarily feature a nice bit of salmon. Like tuna from tuna-fattening pens, salmon is a luxury product. That kind of aquaculture, says Bubba Cook of WWF, is 'sending fish to white tablecloths and restaurants. It's not going to accomplish anything with respect to hunger or feeding the world. It's about making money.'

In Norway, I arranged to meet a man named Terje off the ferry. My heart had sunk at the thought of yet another long schlep through Norway's public transport system to its edgeland, this time to Krokeide, a place that sounds like a biscuit. Terje is a marine biologist at the Institute of Marine Research on the island of Austevoll, a decent drive away but a short boat ride. Terje looks like a central casting sailor with his sailing cap and his white beard. He could definitely advertise fish fingers. We will get the boat to the lab, he says, because the road is closed. The surroundings are dramatic: slate-grey water, snow-covered islands. It is so dramatic that during the pandemic when people were sharing pictures of their home offices, Terje's wife Grø thought she couldn't share hers because it wouldn't be fair.

The Institute of Marine Research is notable in Norway. It houses academics and a research station. Terje phones to ask for permission for me to visit the labs: I can visit some bits but not the brood stock halls, because fish fry are easily spooked. The lab is a short walk away through thick snow – it has been falling all morning, again – to a low complex on the water.

On a door a sign reads 'PLEASE KEEP THIS SHUT TO KEEP OUT OUR FOUR-LEGGED FRIENDS'.

'Dogs?'

'Otters. Mink. Fish-stealers.'

In this area there is a tank of tiny plaice that skitter when we approach, then ballan wrasse, blue and brown wrasse, then cod eggs. On a notice board, someone has drawn a pretty good cartoon of a lumpfish. Terje's specialty is feeding fish, but not with pellets. He works with copepods, artemia, and rotifers; tiny marine animals. (My best dinner-table artemia fact: a tenth-century Iranian writer named Estakhri referred to this tiny crustacean as an 'aquatic dog', and nobody knows why. It looks nothing like a dog.) Copepods are best because they contain all the right phospholipids. But they are hard and slow to grow, and so for decades people have been trying to mimic the profile of copepods with ingenuity and patience, feeding omega-3 and iodine to artemia.

Terje has worked on how best to feed salmon, but also cod. Cod, he says, is 'the new boom'. Actually, it's the second boom of the new boom. The first failed because cod farming was too difficult. There were massive escapes: cod were biting through the nets. They were eating each other. The industry collapsed. But the glitter of the salmon riches must have dazzled and 10 years ago cod farming came back, although it is so far limited. The new boom consists mostly of three farms near Trondheim run by Norcod, a company that began selling its product in 2021 after four years of trials.[40]

Terje tells me of a cod that was tagged and in a tank with a knob in the middle that released food. His tag got stuck on the knob and he realized he could get food that way and did, repeatedly. 'Clever animals.' He also says you can spot a farmed cod by its deformities: because they swim round and round in a tank, their heads get bent. Terje went up to Bodø recently to do some cod detecting. Fishermen were catching deformed cod and thought they were from the nearby farms, but the Norcod owner denied he had had any escapes. The fishermen were calling it 'monster cod'. Terje checked the cod's genetics and proved Norcod wrong. At Easter, a skipper fished a strange fish off the Lofoten Islands in the north. The cod's head was deformed: 'There was something strange about the neck and the liver was snow-white.'

Collected and tested by the Norwegian Directorate of Fisheries, the cod was identified as a farmed one. Norcod's farmed cod sells for 8 per cent more than wild-caught cod, according to its CEO. He calls it 'snow cod: the new saltwater superstar'.

The Institute's fish-rearing experiments have a broad reach. They have reared plaice, halibut, and cod, but they have also tried eels. Never again. They died. 'The Japanese have tried to farm them, but you have to feed them shark egg paste four times a day. It's very complicated.' Halibut are a better prospect, but Terje is betting on lemon sole and plaice. If you were going to pick a fish that was perfect to be farmed, it would not be a salmon. They are carnivores and need to be fed so carefully; there are entire university departments figuring out salmon-friendly ingredients.

Sea cucumbers are on the easier side of aquaculture, along with mussels: you stick them in simple pens on the seabed and wait for nine months. Norway once attempted to farm wolffish, an ideal fish to grow according to the Global Seafood Alliance, as it has 'high specific growth rates in captivity at very high densities, a high yield of 0.7 kg to 1.1 kg fillets, nonaggressive behaviour and few disease problems'. But that failed in 2007 when the only company to attempt it killed half the stock when a water pump broke. Also, chefs don't know what to do with it. Customer palates are a barrier when it comes to aquaculture.

Leave out Norway, and salmon is not even in the top five of the most farmed aquatic products in Europe. The top seven are mussels, trout, oysters, seabream, seabass, carp, and tuna.[41] Numbers of how many species of fishes and shellfish are farmed vary from 100 to more than 400. Minister Someone Who Does No Harm should leave plenty of room on his aquaculture plate for (to use the FAO's list of major farmed species) grass and silver carp, Wuchang bream, and largemouth black bass, or (at sea) Japanese amberjack, European seabass, milkfish, and barramundi. He, and all of us, should be investing in land-based aquaculture, freshwater aquaculture, anything but marine aquaculture of carnivorous highly migratory fishes.

In China, fishes have been farmed for 4,000 years. Today China is the giant of aquaculture and calls its industry 'Blue Granary'.[42] Although sea salmon farms get attention, globally the majority of fish farming is freshwater. Some is done on land and more should be done on land, in 'closed containment systems'. No sea lice, no escapes, no damaging wild fish populations. It can be expensive, but it can also, I read in one paper, 'be done in a warehouse in Glasgow'. It needs imagination, just like that displayed by China's frog–rice–fish farms. When fish are introduced into rice farms, rice yield improves. Fish and frogs are even better. The rice provides shade for the animals, and the animals can eat the insects that fall off the rice. Also the fish and frog shit gives rice valuable nutrients: this is important when many soils are deficient in phosphorus. Frogs are also less trouble than ducks, which tend to gobble the rice ears.[43]

In 2020, China produced 20.6 million tilapia, a fish known as 'aquatic chicken' for being cheap and tasting good.[44] Unlike the finicky salmon, tilapia resist disease better and don't mind poor water quality. But tilapia doesn't have EHA or DHA, essential acids that many consumers eat fish to acquire. In fact, the best sustainable farmed marine creature is a mussel or an oyster. They don't need feeding, they are easy to grow, and they filter the water. The day that Tony the fishmonger's customers, and all of us, ask for 'a nice bit of mussel' for tea will be a better one for salmon, and for the ocean.

Middlegrunden wind farm off Øresund

14

Windy City

Kites rise highest against the wind, not with it.

UNKNOWN

More than 80 per cent of the oceans has never been mapped, observed, or explored.[1] That does not mean that we have left them alone. Fishing industrialized the ocean in the 1950s and has not stopped. A trip to the Indian Ocean at night will reveal a mass of boats, lights blazing to attract squid. With sights like that it is hard to believe the ocean is unsullied or empty. Consider too all the static structures like oil rigs, pipes, cables, and offshore wind turbines.

Oil rigs should be our past but offshore wind is meant to be our future. Any offshore wind site will generally achieve two things: it will produce electricity that is not derived from fossil fuels and it will infuriate fishermen. The conflict between fishermen who think their fishing grounds will be ruined and the offshore wind installers who don't is called a spatial squeeze, even in many square miles of ocean. Fisheries conflicts are not new. Between the end of the Second World War and the late 1990s, a quarter of militarized disputes between democracies have been over fisheries.[2] In 2023, the most interesting space being spatially squeezed was the Atlantic Ocean off the Eastern Seaboard of the United States.

I have accidentally timed my arrival in New Bedford, Massachusetts with a nor'easter, a poetic-sounding wind that has dumped a foot of snow on the city in late February, when the city thought it was done with snow. It is too cold and icy to walk, so John Regan – communications chief for the Port of New Bedford – and I sit in his warm SUV to tour 'the most important port in America by value of catch', according to its (not very catchy) branding. The tour doesn't take long. New Bedford is a vibrant and lucrative port that generates $11.1 billion a year, but it is not huge.[3] In fact, the problem facing the port authority is where to fit everybody. Already fishing boats are double-parked, tied two or three abreast, because there isn't enough wharf space. And now they have to share that space with the offshore wind people.

On the other side of the Atlantic, offshore wind has been thriving for decades. The UK has 43 operational offshore wind farms (the industry prefers 'arrays') that generate 17 per cent of the country's electricity.[4] With his usual errant diplomacy, the former prime minister Boris Johnson told the UN that the UK would be the 'Saudi Arabia of wind'.[5] Until 2023, the total number of operational wind turbine generators off the United States was five.[6] But then, as part of President Joe Biden's 2021 climate change plan, offshore wind was hoisted up the priority list.[7] By 2030, offshore wind was supposed to produce 30 gigawatts of energy and eventually 10 per cent of the US's energy.[8] The Bureau of Ocean Energy was to review 12 wind leases for offshore sites over the subsequent five years.[9] For now, only Vineyard Wind, a Danish-owned company, has leased land in the Port of New Bedford, but another has signed a lease, and others are waiting. The subject headings of the city newspaper include 'Fishing' and the 'Daily Catch' but also 'Wind'. If there is a Wild East of wind, New Bedford is it.

It makes sense. Although Vineyard Wind's first array is 30 miles from New Bedford, the port has better facilities than others that are closer: space, electricity, and access. Wind's footprint is evident all over the port: the gigantic turbine blades, the massive yet stumpy foundation blocks, the barges that carry everything out to sea, squeezing through New Bedford's gap in its hurricane barrier with such tightness that each transit is carefully monitored by the port staff and usually has at least three tugboats standing by just in case.

John Regan is as warm and charming in life as he is economical in written communication. His Dartmouth accent helps too: that never sounds cold, with its long 'aaaa' and words that take a peculiar turn now and then. Regan is one of only six staff members running this port, and his corporate headaches are only going to get deeper and wider. Not just because of that awkward hurricane barrier gap. The port infrastructure is so old that bits of the wharves keep dropping off. They will have to close a whole wharf because they want to raise it and make it bigger. 'Where will you put the boats?' 'We don't know yet. We have a year.'

New Bedford has history at being a successful port. In the past its biggest talent was whaling. It is proud of that fact and has a lavish whaling museum that does not address the ethics of killing all those whales, not once, and there are Moby Dick references all over town. There is a Whale Tooth's Parking Lot but, although the scallop fishery generates most of the port's $300 million or so in fishing revenue, I never find a Bivalve Alley or a Scallop Street.[10] New Bedford has changed and now it will change again.

That is because of geography and wind. What offshore wind people want is wind but also an ocean shallow enough for them to sink turbines into. These two factors are hard to find together but the Atlantic Ocean's Outer Continental Shelf delivers both. In that it resembles the North Sea, which now has hundreds of wind turbine arrays. Of the dozens of companies that have secured leases to construct arrays, Vineyard Wind was the second to generate any electricity, after the Block Island array off Rhode Island, whose five turbines started production in 2016. In 2004, the company Cape Wind wanted to erect 130 wind turbines 5 miles off Martha's Vineyard and 15 miles from Nantucket. The company withdrew its plans in 2017 because of objections from homeowners who thought their view would be spoiled.[11] That some of those homeowners were senators may have been a factor. Andrew Minkiewicz is now a lawyer who works with the fishing industry on the East Coast, but he used to be a staffer on Capitol Hill. 'I was directed to kill that project.' When the wind farm people came, he remembers thinking, 'Good luck,' because he knew who lived nearby.

This time round, the wind companies were prepared. Most are foreign. The Danes are big in wind and, more importantly, in wind farm infrastructure; the United States has no suitable manufacturers. Most of the companies have employed fisheries liaison managers to deal with the most obvious opposition. There are no powerful senators living on the coast at New Bedford, where more than 20 per cent of people live below the poverty line, twice the rate of Massachusetts as a whole.[12] But there are fishermen and a flourishing scallop industry and a whole sea of possible hostility.

Crista Banks is the response to that. She works now as fisheries liaison manager for Vineyard Wind. Her son is named after a wind, but that is a coincidence because he was born long before she came here, when she was still a fisheries scientist specializing in monkfish.

We meet in Vineyard Wind's office in the Bank of America building. It is on a floor high enough to give a good view of Vineyard's staging area on the docks. I see the turbines lying in wait to be transported and the stumpy foundation posts. I ask questions and if Crista doesn't know the answer, she asks someone who does or who has Google.

For example:

Do turbines always turn in the same direction? (Yes, but they can be reoriented 180° to catch the wind.)

How tall is the tower? (850 feet, about the height of the Statue of Liberty, the usual comparison they use.)

How long are the turbine blades? (One and a half football fields.)

That measurement again. I remember it from shipping. I've never heard of a ship that was not described to the public without it being compared to a football field. So the turbine is the same length as the container ship I travelled on. In shipping now that is small, but if you're a fisherman then anything to do with a wind turbine generator is big. It is too big even in all that ocean.

Crista doesn't walk around town wearing Vineyard Wind clothing anymore. She says that's not because she fears aggression, but just because she keeps getting stopped and asked questions. It's for a quieter life. John Regan tells me that Vineyard Wind had to remove posters at the port because they kept getting defaced. The situation is not as tense as in France, where fishermen sailed out to the Saint-Brieuc wind farm off Cherbourg in 2021 and set off distress flares,[13] but there is protest. From the fishermen, but also from unexpected quarters.

Between 2016 and 2022, 175 humpback whales were found dead.[14] That was pre-wind. But when two dead humpbacks were found ashore a week after Vineyard Wind began construction,[15] followed by a highly endangered North Atlantic right whale,[16] convictions began to be formed. Wind kills whales. Wind is bad. US media began talking of a 'clean energy culture war', in which anti-wind and pro-fossil fuel interests found a cause in saving the whales. Soon whale biologists were getting serious threats because they were – obviously – part of a conspiracy to cover up the 'real' cause of whale deaths. What is that? In the words of Donald Trump, it is 'windmills' that are making whales 'crazy' and 'a little batty'.[17] And also a little dead.

In fact, said whale biologist Mark Baumgartner, 'Of all the right whales that have died over the last several decades, the cause of death for juveniles and adults is always vessel strike or fishing gear entanglement.'[18] Warming ocean waters mean fishes are moving, and humpbacks follow. They end up in areas where they are not expected or protected and where human activity has increased. Of 90 humpback whales autopsied by NOAA since 2016, nearly half were linked to vessel strikes or fishing gear entanglement.[19]

The anti-wind whale defenders blame seismic surveys. They demand necropsies of whales' ears. We are certainly deafening the ocean in general with our noise: our ship propellers, our surveys, and our clamour and industry. But for Erica Staaterman, a bio-acoustician at the US government's Bureau of Ocean Energy Management, 'There's a pretty big difference between some of the sound sources used in the oil and gas industry compared to what's typically used in site characterization for offshore wind.' Oil and gas surveys use seismic air guns. They fire into the sea floor. They are loud and violent. Wind farm surveys 'are called high-resolution geophysical sources, and they're typically smaller in the amount of acoustic energy they put into the water column'.[20] Also, they are usually used in millisecond bursts between periods of quiet.

This sounds persuasive. But all human activity in the ocean affects its creatures, even benign renewable energy. In 2023, NOAA and the Responsible Offshore Development Alliance, a fisheries organization, published a 400-page Synthesis of Science.[21] The synthesis was remarkable for two phrases frequently repeated: 'data gap' and 'knowledge gap'. The report considered the impact of electromagnetic fields (EMF) on fish populations. Wind turbines need power cables; power cables emit EMF. Plenty of fishes are either magnetoreceptive or electroreceptive. Either way, they could be disturbed or deranged by EMF, but 'there are presently no thresholds indicating acceptable or unacceptable levels of EMF emissions in the marine environment'.

Vineyard Wind has tried to head off objections with diplomacy. When the spatial plans were first laid out, the array was set out in a north-west to south-east orientation because that was the direction of the fishing traffic. Vineyard planned to have between 0.6 and 1 nautical mile between the turbines. But every other wind company planned its own layout as it pleased. In her personal view, Crista says over lunch, 'It was a shitshow.' The fishing industry objected – half a mile wasn't a safe distance between turbines – and she agreed with them. 'The 1 × 1 nautical mile layout was the best thing to come out of the pushback from the fishing industry because [otherwise] it would have been really hard to navigate through.'

A nautical mile sounds like plenty of room to get a scallop boat through. 'No,' says Eric Hansen. 'One-mile spacing sounds good. It's one mile, it's a long way. But in reality, if you're in the middle of it, you've got a turbine half a mile on one side, half a mile on the other, and half a mile in front of you, and half a mile behind you. And offshore with a 20-foot sea and a rolling boat, that's not very far.' Elsewhere the prospect of fishing among the planned 62 turbines has been compared to flying a kite in a forest.[22]

Hansen had been introduced to me as a 'reasonable voice' in the fishing industry. He has been fishing for scallops for 30 years and his name is known in New Bedford fishing. His grandfather was known as the Terrible Swede and is listed among all the nicknames on a board at the Fishing Heritage Center. I read some out to him. The Irish Midget, the mad Dane, the mad Russian. The Viking. He nods. 'Well, that could be anybody.' In New Bedford, more than half the population claim Portuguese roots and a sizeable minority are descended from Norwegians; the Portuguese came to work in whaling, then the Norwegians came to fish. Now the fishing is ethnically divided: Norwegians do scallops and Portuguese do groundfish. Hansen's son is out this minute on one of his two scallop boats but on a scientific research trip. The scallop industry sets aside a fixed amount of catch to pay for scientific research, usually amounting to about $5 million a year, because, as Hansen remembers, it learned the hard way. In the old days the scallop industry wasn't regulated. He calls it 'unfettered'. 'We were fishing it to extinction with the gear we had. We were retaining everything.' Something had to be done. The ring size on the dredges was changed to four inches, meaning smaller, younger scallops went free, and the fishery lived to be fished. NOAA describes the New England scallop industry as 'a fishery success story'.[23]

Today the industry is so heavily regulated, boats are tied up for most of the year because they have so few days fishing. They can still make a good living because the scallop industry is usually buoyant (not so much this year; when China banned all Japanese seafood because of radiation concerns, Japan began sending its scallops to the United States and the competition has been formidable).[24] It is a living well earned. I meet another wind company manager whose brother is a scalloper and ask if his brother likes it. 'Nobody does. They do it for the money. And this is one of the few jobs where you really earn it: it is brutal, brutal work.'

Scallopers spend hours 'shucking', prising open the shell with a blunt knife and removing the meat. They often get 'the grip', a repetitive stress injury that turns a hand into a claw. There is no remedy at sea except medication. Occupational health people tell me opioid use in fishing is rising rapidly because fishermen take strong drugs to deal with injuries at sea then need it ashore and always. Eric Hansen doesn't get the grip anymore because he rarely goes fishing: he limits himself to one trip a year but only when his scallop boats are working as research vessels. I ask him what the grip feels like. 'Is it painful or do you have restricted movement?' 'All of the above.' The best remedy is to stop shucking. 'I was home for about four years, then I went out shucking scallops, and my wrists had gone. No word of a lie, my whole forearm looked like a pumpkin.'

Hansen sits on boards more than boats these days. His measured way of speaking would be a benefit in any committee meeting. Don't mistake his quietness for passivity. 'The fishing industry gives input,' he says, 'and it goes in one ear and out the other.' Hansen has been involved in discussions about offshore wind since 2014 because he is in a fisherman's working group on offshore wind energy. Even with all the liaising, mistakes have been made that have infuriated fishermen, such as the pilings. When a wind turbine generator is constructed, first there are the surveys. Then the piling is driven into the seabed. But a piling sits only three metres above the water level. It has no lights and is a hazard. 'We asked them to put lights or AIS on them. But they're awash and you can't put lights on something awash. Once it's driven, they can't get on it, it's too dangerous.' An AIS though could be done virtually. You only need one transmitter station. Hansen says that requires a Federal Communications Commission permit, which takes six months, and Vineyard Wind has only just applied for one. 'So in the meantime you've got hazards out there.'

The speed of offshore wind installation may seem rapid, but the current industry dates to 2017, when Vineyard Wind won its first lease. I ask Hansen if his feelings have changed since 2014. He pauses to think. 'I'd have to say yes. I was more optimistic in 2014.' This is surprising: the scallopers were successful at getting wind companies to change locations so as to avoid prime scallop beds. Millions of acres were lopped off the original areas. Now, scallop grounds are barely affected and it's the squid guys who are in trouble. Already, some complain that their nets have been wrecked on Block Island's seabed electricity cables, and a wrecked net can cost $10,000 to repair. Still, New England doesn't have it as bad as Maine: there, a project intended to install floating turbines secured to the seabed with guy ropes. Who can fish around those?

Maybe the New Bedford fishers have the best of a bad job lot. Hansen is content with the revised spacing of Vineyard One's turbines, even though fishermen wanted transit corridors two or three miles wide. He knows things could be worse because he went on a fact-finding trip to the UK, and they did a boat trip out to the Thanet wind farm off the Kent coast. One hundred turbines, seven miles offshore. 'They were like a half mile apart,' says Hansen. 'Claustrophobic.'

In Ramsgate, near Thanet, he was introduced to a fisherman sorting his catch. His British hosts explained to the fisherman that these visiting Americans wanted to get the man's perspective on wind farms because they were being constructed now in the United States. 'He looked us directly in the eye, he says, "Gentlemen, you're fucked." I don't use that language lightly, but that was a direct quote. That hit hard.' A study by the Plymouth Marine Laboratory found that nearly two-thirds of UK fishermen interviewed reported a negative outcome on catches and profitability after turbines were constructed. Their complaints: they couldn't tow gear between the turbines; they had to steam further to find fish. This cost them money and fuel. The respondents believed that nearly 80 per cent of the wind farms operational or under construction have negatively affected fishing activity.[25]

For Crista Banks at Vineyard Wind, the situation is made tougher by data, because there isn't enough. As a scientist, she is frustrated. Each wind farm company has set aside money for scientific research, so she can commission surveys. But then bodies such as NOAA and the Bureau of Ocean Energy Management do not accept the data because it comes from a private source. The 'knowledge gaps' in the NOAA Synthesis of Science apply to the whole industry and its impact on the ocean. Most construction should include a baseline survey, where the environment is assessed before any activity so there is data to assess whether later anything has changed. There is no baseline study for the East Coast wind areas.

Hansen is also frustrated. Without a baseline study, any claims that fishing has been affected are rudderless. 'The fisherman says, "We can't prove that that is what we fish because you don't have the documentation and now the fish aren't there anymore. We can't prove that the fish aren't there anymore because we can't prove that they were ever there in the first place."'

What is known is that little is known. 'Whether it's fishermen losing their fishing grounds,' says Crista, 'or fish species changing where they live. Even if certain species are attracted to our wind turbines and are there, that can be great, but if management doesn't allow the fishermen to catch them, just like you're seeing off England, then it's still not going to help them. So we're just throwing a shitload of uncertainty into an already stressed moving body.' It is established that fishes like static structures. They can hide and breed and hang out on them. They might do this on the wind turbines; they might not.

Currently the only certainty about America's wind revolution is that the turbines will spin clockwise because they always do. Crista is now wind but she hasn't forgotten fisheries. She says, 'I don't know any fisherman who is in favour of offshore wind,' and she is including those who work for her company. 'They say to me all the time, we'll work with you but if your company packs up and goes under, that would be great too.' She says, 'I'll take that, because everything about offshore wind will and can potentially impact the fishermen and their ability to make money.'

Vineyard Wind has not yet packed up and gone under, though several companies have withdrawn their leases. The problem[26] is money and whale poo. Glauconite is a substance (one newspaper calls it 'a tricky, sticky substance') formed from fossilized whale shit that crushes under high pressure and turns into something like clay when it is exposed to friction. So, pile-driving into it means getting what the wind industry calls 'pile refusal'. One solution is a 'suction bucket', a shallower foundation that, unlike deeper pile-driven bases, shouldn't penetrate unstable glauconite. Fishermen probably won't like that either: already they have to deal with hazards caused by 'scour', rocks that are arranged around the base of piles for stability.

Eric Hansen is not against wind energy. 'I'm not a climate change denier at all. I think green energy is probably our future. But seafood is also a renewable resource. And it's a food for the nation.' (I don't point out that the United States exports most of its fish; he knows that.) 'The only analogy that I can give you is we know there's a bunch of steamrollers that are coming down the road and if we don't redirect them, at some point they're going to run right over us.'

Before the fishing

15

Fish In, Fish Out

All through the oceans there are ghosts. Fishing vessels lose so much gear that fishermen budget for it. This lost gear becomes ghost nets, ghost pots, ghost lines, and ghost hooks. It contributes 10 per cent of ocean plastic, which is bad. And it keeps on fishing. Fred Nunn is secretary of Ghost Fishing, an English group of volunteer divers that tries to collect it. WWF calls ghost gear 'the most deadly form of marine plastic debris'. Ghost gear harms two-thirds of sea mammals, half of seabirds, and all the turtles. A single lost or abandoned net, wrote Ingrid Giskes of the Global Ghost Gear Initiative, 'is estimated to kill an average of 500,000 marine invertebrates (think crabs and shrimp), 1,700 fish and four seabirds'.[1] Some more big numbers: researchers last year interviewed 451 fishermen from seven countries about how much gear they lost, then multiplied the responses by global fishing effort data. This came to 1,864 miles of gill nets, 47,000 square miles of purse seine nets, 135 square miles of trawl nets, 459,618 miles of longline mainlines, and more than 25 million pots and traps. Against all that, the 70 volunteers of Ghost Fishing may not seem like much. But they have vim and they can dive. And they are not alone. The first fishing trash collectors in the UK were Neptune's Army, based in Wales, and there are other small bands around the world. 'People who sign up now,' says Fred, 'the story seems to be the same for everybody: "I'm fed up of seeing this in my local dive site, I don't like to see the suffering of creatures, I want to do something about it."'

At first, they got no welcome from the fishing industry. 'As soon as you say you're an eco group, or as soon as you say, we're coming to remove this, the typical response from fishermen is, "We don't do ghost fishing. We don't lose any equipment overboard."' In fact, fishermen who did eventually talk to them told them they allot 10–15 per cent of their annual budget for replacing lost and damaged equipment. 'And they are not necessarily bringing all that equipment ashore or paying to have it disposed of. So we know it's getting lost.' Ghost Fishing does not want to blame the fishing industry. It wants to help them out. 'The ideal scenario is they get something lost or snagged up, they can contact us, we can go and do something about it and return it to them in a usable state.' A new creel pot can cost £100 ($125).

Then the tide turned. Younger fishermen saw the divers bringing in pots and wanted to know where they were from. 'We would say, "We've recovered them," and they would say, "Do you want them?"' The pots are fair game. Fishing gear is rarely marked or traceable. Requiring both these things would make lost gear easier to deal with. Making gear more biodegradable would be even better. If a net dissolves in water, it can't strangle a seal.

Flush with their free pots, fishermen were soon telling divers where to go to find stuff. 'I lost a string over here and I know someone lost a string over there.' In Falmouth they found a string of 36 pots half a mile long. Fred had seen it on a dive, and they had gone back to find it, but there was no identifier to help them. No useful wreck, no buoys. The divers split up and swam in opposite directions, and they were each swimming for 40 minutes before they found the ends. The boat skipper was phoning local fishermen. 'Is it yours?' There was still fresh catch in the pots. There's no reason a pot won't go on catching: the catch dies and becomes bait, then more catch arrives. Fred calls this 'the ghost fishing cycle'. 'We've got proven evidence that it happens in pots, which some fishing associations deny is possible. They say crabs will cut themselves out after a couple of months. But we've never seen any evidence of crustaceans cutting or eating their way out. So we're calling BS on that.' In Scandinavia, pots must have a biodegradable panel so that catch can escape. Ghost Fishing would like to campaign for that same rule or law in the UK, but they are already strapped for cash and dealing with volunteer burnout. Even so, they went to a skipper show in Aberdeen, a trade show for fishing gear. 'Some of the fishermen came up to us and it was the usual, "Are you fishing for ghosts?" Then once we got chatting, they'd say, oh you should come to our area. There's definitely gear there.' Ghost Fishing has retrieved parts of gill nets that are 300 metres long. Those are the smaller kind, from inshore vessels. Fred calls gill nets 'a double-edged sword'. They are meant to be the kindest nets because they catch selectively according to the size of the mesh. 'But once it's lost, it will continue to catch. Scavengers will come and they will get caught too. When they're used properly, they are sustainable. Better than the big purse seine nets that they just drag through the sea and hoover up everything. Gill nets are what we really want the fishermen to report to us as soon as possible, and we'll put together a quick response time. Because we find real devastation in those.'

We are so powerful that we even kill fish without wanting to. There is a lot wrong with that. There are things wrong with the fishing industry in general, but what seems deeply complicated is also simple. Fisheries scientists talk of 'fishing pressure', and we must ease it. We need brakes. Where will we get them and who will apply them? In March 2023, yet another international treaty was formalized after a final 36-hour debating session.[2] This one is a magnificent thing and praise of it has been high and loud. Much of it has included the word 'finally'. Greenpeace thinks it 'the biggest conservation measure ever'.[3]

The treaty's full name is 'The Agreement under the United Nations Convention on the Law of the Sea on the Conservation and Sustainable Use of Marine Biological Diversity of Areas beyond National Jurisdiction', which is supposed to be shortened to BBNJ, for 'Biodiversity Beyond National Jurisdiction'. Most people call it the Global Ocean Treaty or High Seas Treaty. Whatever its name, it is good news. The ocean is divided into coastal waters and exclusive economic zones.

EEZs belong to states, and states argue about them still. The high seas belong to no state and have been a free-for-all. Or a free-for-anyone with enough fuel and subsidies to get there and take the fish. The Global Ocean Treaty will enable states to turn the high seas into protected areas. Even better, the treaty does not require consensus, a requirement which slows down all international agreements. That means it may actually work. The treaty comes into force when 60 countries both sign and then ratify it by incorporating it into national law. Watchers want this done by the UN Ocean Conference in June 2025. For a while I use ratification trackers set up by NGOs and wait with glee for the numbers to mount.[4] At last count 104 countries have signed, but only 13 have ratified.

The fisheries ecologist Daniel Pauly refers to himself as a 'grumpy old man'. When I ask if he is an optimist or a pessimist, he says, 'This is nonsense. Not a good question. Because it means, can you foresee whether we will win, whether we will save biodiversity, whether we will end up reducing climate change impacts, even reducing emissions? I don't know. I can't foresee the future, but there is only one attitude I can have about it. I'm against it.'

I have another Daniel Pauly quote written down, from a review he wrote of a book about menhaden, another small oily fish – like sardinella – that has been plundered beyond common sense. Pauly wondered whether the book would be 'yet another helpless commentary on how we are trashing our oceans'.

We are not helpless. Better data will help: it will help us understand what fishing boats are actually up to; it will help us understand more precisely how fish are doing; it will help everything. Technology and emerging technology are probably the best hope we have of removing secrecy from the sea. Bubba Cook of WWF, as well as working on tuna fisheries and in the interests of observers, also organizes a conference called Seafood and Fisheries Emerging Technology. It sounds dull but it isn't. His serious manner does not match his nickname ('it's a common nickname in Texas where I'm from and yes I know about *Forrest Gump*') because trying to fix the oceans is a serious matter. The current state of fisheries technology is like the current state of fisheries stock assessments: patchy and unsatisfactory. Global Fishing Watch is doing a remarkable job of using AIS and VMS data to track fishing activity and keeping it accessible. Boats can still switch off or spoof their AIS, and VMS data is often kept hidden by governments. (Global Fishing Watch can still calculate fishing efforts when boats go dark.) A better solution would be to do what development people call a leapfrog. Many Africans without telephones leapfrogged over landlines and went straight to mobiles. Maybe the fishing industry can be made to leapfrog to automatic transfer of data. Cook dreams of boats that use radar to transmit a signal that can't be fiddled with.

When I investigated the world of shipping and wrote that only 2 per cent of containers were physically inspected, my mother said with the wisdom of mothers, 'If people smuggle things in containers, why don't they put windows in them?' The window in fishing is a Unique Vessel Identifier. Cargo ships have this, fishing boats don't. Millions of boats, pirogues, cayocas, draggers, sampans, trollers, and punts can be anonymous at sea. 'It's ridiculous,' says Steve Trent of the Environmental Justice Foundation. 'Every cow in Europe has a passport, and you're telling me we can't put numbers on fishing boats?'

It's not difficult to fix what is wrong with fishing. Many of the distant-water fleets that do the most damage to our fishes and the ocean would be hobbled if they didn't have lavish fuel subsidies from governments. The fishing industry gets $35 billion in subsidies, and $20 billion of that contributes to overfishing. Without it, the greedy boats would have to fish closer to home. They would have to be Rex Harrison of Filey: modest in their appetites. In 2022, the World Trade Organization (WTO) passed an agreement to ban fishing subsidies. It has still to be ratified but if that happens, it will be straightforward for the WTO to request another member to stop subsidizing IUU fishing, fishing over-depleted populations, or engaging in high-seas or distant-water fishing. Critics think the agreement has holes fathoms wide. It doesn't stop 'capacity building' subsidies, so countries can still top up fuel or engine replacement costs, things that enable distant-water fishing in the first place. In short, said the marine conservationist Claire Nouvian, the subsidies 'create incentive to fish too much, too long, and too far'. She called the agreement 'unfinished business'.[5] Unfinished business as usual, full steam ahead.

MPAs should be protected and there should be more of them. Small-scale fishermen should be protected because restraint will keep more fishes in the sea than greed will. When No Take Zones are enforced, fish populations recover. I could go bigger, as plenty of fisheries scientists now do, and wish to close the high seas to fishing. Actually that is not such a big request: only 10 per cent of fishing happens there anyway. But 10 per cent more ocean for fish to rebuild populations would be an enormous result from a modest thing.

Neither the WTO agreement nor the Global Ocean Treaty is perfect. But they mean that things are afoot for the ocean and that is good. I want Rex and Tom to be able to fish, because I have never met a fisherman who doesn't love it and I have never met a retired fisherman who does not miss the sea. (I'd get Tom to switch to pole and line though.)

I would love it to be easier for fish-eaters to make an informed choice about which fish is least harmful to eat. I research online 'Which fish should I eat?' and all the results tell me which choice of fish is healthiest for humans, not which choice of fish is healthiest for the ocean. At the moment, the best option is probably to choose something different. A farmed trout with chips. A bivalve cocktail. Something to relieve the pressure on the most overfished species, like a fish venting device – pop! – but for entire populations.

And if I had one final wish for the fishes and for ourselves, it would be for prudence and patience.

In the last few years, Atlantic bluefin tuna have returned to the seas off south-west England, where they had not been seen for 50 years. People expressed general wonder and pleasure: how magnificent they are! How fast and sleek! What a privilege to see them back in our seas.

As soon as it could, the UK government issued licences for a trial fishery. These impressive creatures cannot be left to be impressive.

Fish In, Fish Out

They have to become 39 tonnes a year of not magnificent fishes but 'catch', although the UK does not need 39 tonnes of bluefin tuna for its food security. This is plunder not prudence.

'We'd have more fish in the oceans,' a marine biologist once told me, as if he was saying the sky was the sky and the ocean was wet, 'if we just stopped fishing so hard.'

Acknowledgements

I thought writing about shipping was complicated but then I chose to write about fishing. I have needed a lot of help with this book from all sorts of places. I hope I remember everyone who should be thanked but if you are missing, my apologies and I thank you anyway. In no particular order, my thanks to fish experts Chris Williams, Laurence Hartwell, Tom McClure, Karen McVeigh, Stephen 'Cod' Astley, Tina Barnes, Tony Walker, and Will McCallum. Stella Maris, last featured in my shipping book, were equally obliging with this one and I thank Bryony and Steve in particular. For their kindness to me and my troublesome stomach at sea, I thank Rex, Alexander, Dave, Cod, and the two Toms.

The writing of this book has coincided with a pandemic (for everyone), long Covid (for me), and some personal upheaval (also for me). I have needed a cocoon of friends, and I thank Katie, Paul, Flo, Sally, Angie, Jo, and Ian in Rivel (as well as Freya the dog); and Louise, Tessa, Griet, Elliot, Amy, Tom, Lisa, and all my fell-running friends who have been my scaffolding and support. I have had good care from therapists and physios along the way, but I am also grateful to Lac de Montbel and Roundhay Lake. There's nothing like sharp cold water to cure everything. You can't cry when you're swimming.

For reading drafts and sections of the manuscript at various stages of writing, I thank Tina Barnes, Jack Clarke, Jerry Dzugan, Elliot Elam, Ruth Metzstein, Tom Ridgway, and Chris Williams.

For the regular images of fish puns on a blackboard ('When a fish swims on by and brushes your thigh, that's a moray'; 'What makes fish

Acknowledgements

terrible journalists? They always spread hake news'), I salute Graham the fishmonger of Tunbridge Wells, and Laura Barber for sending them to me. Laura is a talented fish-pun collector but also my editor at Granta, and I am always in awe of her calmness under fire, her laser focus, and her kindness. She has edited me now for many years and I am very lucky. In the United States, I am also ably looked after by Jill Bialosky and Laura Mucha at W. W. Norton: thank you both. Publishing a book requires a brigade of people and although I may not know their names, I thank all copy editors, proofreaders, production staff, designers, typesetters. You are all essential and I am grateful.

As ever, my agent Erin Malone at William Morris Endeavor has been a rock in a very hard place, sending support and weighted blankets as required. Thank you Erin: we've been together for nearly twenty years and I never stop being grateful for that, as well as to Simon Winchester, who introduced us. Thanks also to Fiona in the London office, to Erin and Fiona's continually efficient assistants, and to all WME staff.

My family – my mother Sheila Wainwright and my brother Simon George – are as families should be: always there. Thank you. You are the best. Or as we say in Yorkshire, you're reight.

All mistakes are mine and I hope mendable and, if not, forgivable.

Which Fish Guide

With thanks to Michael Pollan, an overall guideline: eat fish, not too much, and not the usual.

Most nations have a limited palate when it comes to fish, putting pressure on a small number of fish populations that can no longer bear it. In the UK, most people eat the 'big five' – salmon, tuna, cod, haddock, and prawns – even though 150 species are landed from British waters. Japan's top five are chum salmon, Pacific bluefin tuna, Pacific saury, horse mackerel, and mackerel. What is most likely to be on your plate depends on where your plate is in the world. And how you choose your fish will be informed by your values: Do you want to buy local and not encourage fisheries dependent on fuel subsidies? Do you want to protect fish populations? Do you want to protect the livelihood of local fishermen? All these aims can lead you to different choices. But there is an easy principle to follow: check the most consumed fish where you live and have something else instead. A freshwater trout not salmon; a coley not a cod.

Is the best fish one you can buy from a small-scale fishing boat? Sometimes but not always. The US Food and Drug Administration warns about fish caught by family and friends: you should only eat that once a week because 'larger carp, catfish, trout, and perch are more likely to have fish advisories due to mercury or other contaminants'. Small-scale vessels mostly involve less by-catch but not always. Some large fisheries involving large vessels are deemed sustainable by the most common seafood certification system, the MSC blue tick.

The Marine Conservation Society's 'Good Fish Guide' and Monterey

Bay's 'Seafood Watch' are both useful and detailed. The International Union for Conservation of Nature has a red list of marine species that should be avoided because they are in danger, but the website is not user-friendly and you will have to wade through pages of snails.

The following are a few examples of popular fish choices with some alternatives. They may help for those minutes you stand in front of the supermarket fish fridge trying to eat the right thing. They may not. It's not easy. But any effort is better than none and thank you for making it.

Tuna

Ocean health: Choose pole-and-line caught, troll caught, or free school. Longline fishing catches too much by-catch, so avoid. In general, if the method and the species of tuna are not specified on the package or the can, be wary. Even better, avoid, and demand that the big canned tuna companies get more honest about labelling. 'Dolphin-friendly' may be good for dolphins but not for hundreds of other by-caught species.

Your health: Tuna is a predator and high in the food chain, so it is likely to have absorbed the most amount of heavy metals and microplastics. But it is also extremely nutritious. Pick your species very carefully. Definitely avoid Pacific bluefin tuna: it is now at 10 per cent of its historic population. Smaller tunas – skipjack, albacore – are always a better choice than bluefin.

Alternative? Sardines or anchovies, both high in omega-3s.

Whitefish

Ocean health: Give cod, haddock, and whiting a break: their populations are under extreme stress and need to be allowed to rebuild. If you must have cod, choose from the two cod fisheries that get a blue tick in the MSC guide, both certified Icelandic fisheries in FAO region 27 (Northeast Atlantic). Eighteen other cod fisheries either need improvement or must be avoided altogether.

Your health: Whitefish is low fat and high protein and ticks all the health boxes that you'd hope for: It can reduce blood pressure, keep your heart healthy, and maybe – although the research is not as solid as a clamshell – oil your brain.

Alternative? In the UK, hake or plaice. Alaska pollock is a good choice in the United States and elsewhere, but not pollack from the English Channel or Irish Sea (different spelling, same species), which is red-listed and in a sorry state. All these alternatives are as meaty and flaky as the usual choices. Other options: coley or plaice.

Salmon

Ocean health: Sea-based salmon farms are largely bad for the ocean and for salmon, no matter how well-managed. Farmed salmon is a luxury food and you can easily substitute it for something equally nutritious and probably cheaper. Perhaps in 10 years salmon will be raised in land-based aquaculture, which is better for the ocean (but has its own energy impact). But for now, choose something different when you can.

Your health: Farmed salmon are often treated with chemicals and antibiotics to control common diseases and sea lice. The death toll on salmon farms will not affect your health but it may make you think. Salmon is an oily fish and contains good-for-you protein and omega-3s, and it's easy to cook.

Alternative? Freshwater trout. Farmed is fine.

Tilapia

Ocean health: Tilapia is the most farmed fish in 120 countries. Most tilapia are farmed in freshwater inland. China's tilapia farming industry is regularly criticized for its 'habitat, chemicals, escape, and disease impacts'. Concerns include the use of banned antibiotics and tilapia escaping from flooded farms (tilapia are highly invasive).

Your health: Americans eat half a kilogram of tilapia a year, probably because it is easy to prepare, not too bony, high in protein, and contains good things like vitamins D and B12 and high amounts of omega-3 and omega-6 fatty acids. Seafood Watch recommends checking where the tilapia was farmed. If it was in China or there is no provenance listed, choose something else or tilapia from somewhere else. Colombian, Taiwanese, or Indonesian farmed tilapia is a better choice. Mexican tilapia is rated yellow, which means it is more questionable.

Alternative? Catfish has a mild flavour and a firm texture, like tilapia. The best farmed catfish are grown in indoor recirculating tanks with decent wastewater treatment. Catfish grown in ponds is OK too.

Prawns
Ocean health: Some prawn fisheries entail 90 per cent by-catch. That is unacceptable. A responsibly farmed prawn is better. Who farms prawns responsibly? It depends on the prawn. If it's a king prawn, not Indonesia, India, or Vietnam, according to the MSC. South American farmed prawns are better. If it's tiger prawn, Indonesian is considered a best choice.

Your health: Prawns contain iodine, zinc, and selenium; all good things. They also contain astaxanthin (like wild salmon), which has anti-inflammatory properties.

Alternative? A bivalve. There is no more sustainable seafood than a rope-grown mussel: it doesn't require feeding and it filters the water that it lives in. Mussels are high protein and low fat and contain vitamins A and B12, as well as iron and the usual omega-3s (though they don't have as many as oily fish). But because they filter water, even rope-grown mussels grown away from the sea floor may contain toxins and may not be suitable for certain vulnerable groups.

Still confused? I'm not surprised. Advice goes out of date and depends on where you are and what you are aiming for when you buy fish. So a broad principle may be the most helpful: eat something different, and ideally a mussel or a small oily fish from a sustainable fishery (so not from West African waters, for example). Be adventurous if you can.

Online Guides

The 'Good Fish Guide' by the Marine Conservation Society is the most comprehensive and well researched, and it is regularly updated. English only.
   https://www.mcsuk.org/goodfishguide/

Next, Monterey Bay Aquarium's 'Seafood Watch'. You can search in English, Spanish, or by sushi type. You can also order a printed guide.
   https://www.seafoodwatch.org

Japan is late to the sustainable seafood guide game but now provides the Blue Seafood Guide. Japanese only.
   https://sailorsforthesea.jp/blueseafood

WWF Hong Kong produces a bilingual English/Chinese guide with pretty illustrations. It is aimed at Hong Kong residents but could serve mainland Chinese too.
   https://seafood-guide.wwf.org.hk/en/seafood-guide

Fishbase is an extraordinarily rich resource for everything you want to know about fish biology and fish facts. Set up by Seas Around Us, it can keep you occupied for hours. There's also a fish quiz.
   www.fishbase.se

Local Catch: This network of over 255 seafood businesses aims to support community-supported fisheries in the US, a business model that encourages direct sales between fishermen and consumers. Local Catch's 'Seafood Finder' works even for the landlocked, and you can buy from the retailer or head to a dock.
   https://finder.localcatch.org/

Notes

EPIGRAPH

1. Jim Leape, Mark Abbott, Hide Sakaguchi et al., *Technology, Data and New Models for Sustainably Managing Ocean Resources* (World Resources Institute, 2020).

1. DORA'S BREAKFAST

1. James Chimiak, MD, ed., 'Chapter 1: Basics of Your Heart & Circulatory System', *The Heart & Diving* (Divers Alert Network).
2. Simon Mitchell, MB, ChB, PhD, 'The Physiology of Compressed Gas Diving', Alert Diver Category: Features, Divers Alert Network, 1 August 2016.
3. Art Levy, 'Icon: Eugenie Clark', *Florida Trend*, 2 November 2020.
4. Daniel Pauly, 'Aquacalypse Now', *New Republic*, 7 October 2009.
5. Chris Armstrong, 'Short Cuts: High Seas Fishing', *London Review of Books*, 45, no. 10 (18 May 2023).
6. Helen Scales, 'Why the Once Common European Eel Is Now Critically Endangered (and What Can Be Done about It)', WWT, 14 June 2019.
7. John Wyatt Greenlee, 'Fishing for Gold: How Eels Powered the Mediaeval Economy', *BBC History Extra*, 8 December 2020.
8. National Oceanic and Atmospheric Administration, *Status of Stocks 2023* (May 2024), p.3.
9. Food and Agriculture Organization, *The State of World Fisheries and Aquaculture 2024, Blue Transformation in Action* (FAO, 2024), p.43.
10. Mukhisa Kituyi and Peter Thomson, '90% of Fish Stocks Are Used Up – Fisheries Subsidies Must Stop', UN Trade and Development, 13 July 2018.

11 Environmental Defense Fund, 'What Is Overfishing?' *Overfishing: The Most Serious Threat to Our Oceans*, 18 September 2023.

12 Chris Armstrong, A Blue New Deal: Why We Need a New Politics for the Ocean (Yale University Press, 2022), p.64.

13 International Maritime Organization, 'Enhancing Fishing Vessel Safety to Save Lives'.

14 Graham Edgar, Amanda Bates, Nils Krueck et al., 'Stock Assessment Models Overstate Sustainability of the World's Fisheries', *Science*, 385, no. 6711 (2024), 860–5.

15 Ibid.

16 Armstrong, 'Short Cuts: High Seas Fishing'.

17 Consortium for Wildlife Bycatch Reduction, 'What Is Bycatch?', 2022.

18 Food and Agriculture Organization of the United Nations/Globefish, 'Global Fish Economy: Production and Trade to Grow in 2022, Prices Remain Strong Overall', 17 March 2023.

19 Alison Mood, *Worse Things Happen at Sea: The Welfare of Wild-Caught Fish*, fishcount.org, 2010.

20 Ferris Jabr, 'Fish Feel Pain, Now What?', *Hakai Magazine*, 2 January 2018.

21 Alison Mood and Phil Brooke, 'Estimating Global Numbers of Fishes Caught from the Wild Annually from 2000 to 2019', *Animal Welfare*, 33, e6 (8 February 2024), 1–19.

22 Numbers are fishy, so most documents prefer to use tonnes. The FAO's 2022 report *The State of World Fisheries and Aquaculture* calculates that 178 million tonnes of aquatic animals were raised in aquaculture in 2020. They have not counted individuals.

23 Data is imperfect, but a good analysis of available data is here: https://faunalytics.org/global-animal-slaughter-statistics-and-charts/.

24 Data Genetics, 'One Trillion Dollars'.

25 Stop Illegal Fishing, 'How Much Illegally Caught Seafood Do We Eat? A Fact-Check of the *New York Times*', 14 July 2016.

26 Stephen Leahy, 'Revealed: Seafood Fraud Happening on a Vast Global Scale', *Guardian*, 15 March 2021.

27 Donna-Mareé Cawthorn, Baillie Charles, and Mariani Stefano, 'Generic Names and Mislabeling Conceal High Species Diversity in Global Fisheries Markets', *Conservation Letters* 11 (2018), 22 June 2018.

Notes

28 Pew, 'Despite Progress, Illegal Catch Continues to Reach the Market', 2 August 2023.

29 This figure varies according to trade flows, which is why most government websites talk of the UK importing 'the vast majority' of the seafood it eats. WWF calculated that the UK imported 81 per cent of its consumed seafood in 2019. WWF, 'Risky Seafood Business, Summary Report 2022', p.5.

30 Oppian, *Halieutica* (Digital Loeb Classical Library, 1928).

31 'Possibly one of the best adventures of your life.' https://tuna-tour.com/en/.

32 European Food Safety Authority, 'Species-specific welfare aspects of the main systems of stunning and killing of farmed tuna,' *EFSA Journal*, 1072 (2009), 1–53

33 Felicity Cloake, 'Fish Fingers Turn 60: How Britain Fell for Not-Very-Fishy Sticks of Frozen Protein', *Guardian*, 15 September 2015.

34 Leathercraft Masterclass, 'Stingray Skin: The Most Unforgiving of the Exotics', 10 July 2021.

35 Food and Agriculture Organization of the United Nations, 'The State of World Fisheries and Aquaculture: Blue Transformation in Action' (FAO, 2024), p.xix.

36 Food and Agriculture Organization of the United Nations, 'The State of World Fisheries and Aquaculture: Towards Blue Transformation' (FAO, 2022), p.83.

37 Kawamoto Daigo, 'The High Price of Japanese Seafood: Can Consumption Climb in 2023?', *Nippon.com*, 13 February 2023.

38 Marine Stewardship Council, 'What Sustainable Seafood Species Can You Eat in the UK?'

39 National Oceanographic and Atmospheric Administration, 'Behind the Scenes of the Most Consumed Seafood'.

40 Jon Henley, 'Netherlands Offers Free Herring as Covid Jab Incentive', *Guardian*, 17 June 2021.

41 Agence France Presse, 'Taiwan Official Urges People to Stop Changing their Name to "Salmon"', *Guardian*, 18 March 2021.

42 Oceana, 'Save the Oceans, Feed the World'.

43 Adam Vaughan, 'Global Demand for Fish Expected to Almost Double by 2050', *New Scientist*, 15 September 2021.

44 Victoria Braithwaite, *Do Fish Feel Pain?* (Oxford University Press, 2010).

45 United Kingdom, HM Government, Animal Welfare (Sentience) Act 2022.

46 Jonathan Safran Foer, *Eating Animals* (Penguin Books, 2010), p.31.

47 Angela Nicoletti, 'Sharks Get By with a Little Help from their Friends', University of Florida, 12 August 2020.

48 Daniel Holt and Carole Johnston, 'Evidence of the Lombard Effect in Fishes', *Behavioral Ecology*, 25, no. 4 (July–August 2014), 819–26.

49 Rob Williams, Christine Erbe, I. Made Iwan Dewantama, and I. Gede Hendrawan, 'Effect on ocean noise: Nyepi, a Balinese day of silence', *Oceanography*, 31, no. 2 (2018), 16–18.

50 Cédric Pene and Xiaolu Zhu, '"Agricultural Products" and "Fishery Products" in the GATT and WTO: A History of Relevant Discussions on Product Scope during Negotiations', World Trade Organization working paper, March 2021.

51 Samuel Coleridge, *Rime of the Ancient Mariner*, Text of 1834, Poetry Foundation.

52 Steven Adolf, 'The Bluefin Tuna Paradox', *Tuna Wars*, 13 April 2020.

53 Scarlett R. Howard and Adrian G. Dyer, 'How to Engage Public Support to Protect Overlooked Species', *Animal Sentience*, 27, no. 25 (2020).

54 Almond Alliance of California v. Fish and Game Association, Super. Ct. No. 34201980003216CUWMGDS), C093542, 31 May 2022.

55 Oppian, *Halieutica*.

56 Ibid., p.203.

57 Ibid., p.233.

58 Ibid., p.425.

59 Ibid., p.429.

60 Ibid., p.353.

61 One estimate is 38 million. The Food and Agriculture Organization of the United Nations thinks there are 59.7 million people working in fisheries and aquaculture, and 4.5 million vessels.

62 U. Rashid Sumaila, Andrea Pierruci, Muhammed Oyinlola et al., 'Aquaculture Over-Optimism?', *Frontiers in Marine Science*, 9 (November 2022).

63 Ibid.

64 Oceanographic staff, 'First Marine Fish Declared Extinct', *Oceanographic*, 2 January 2024.

65 Armstrong, 'Short Cuts: High Seas Fishing'.

Notes

2. A MAN IN HIS BOAT

1. 'Why Filey's Fishermen Fear They May Be the Last Generation to Cast Their Nets', *Yorkshire Post*, 3 March 2018.
2. John Worrall, 'The Coastal PO', *Fishing News*, 5 July 2017.
3. '1000-Year History to Save', *Gazette and Herald*, 17 May 2001.
4. Irene Allen and Andrew Todd, *Filey: A Yorkshire Fishing Town* (Allen and Todd, 1987), p.6.
5. Ibid., p.4.
6. Louise Roddon, 'The 6 Seaside Towns Everyone's Going to Be Talking About This Summer', *Sunday Times*, 22 June 2022.
7. Allen and Todd, *Filey: A Yorkshire Fishing Town*, p.37.
8. UK Government, Sea Fisheries Commission 1863, 'Evidence from Filey', 1 October 1863.
9. Allen and Todd, *Filey: A Yorkshire Fishing Town*, p.67.
10. The comparison with lawnmowers has been made often, e.g. in https://www.bloomberg.com/features/2023-brexit-fails-uk-fishing/ The UK fishing industry in 2023 landed catch worth £1.1 billion. Valuation of the UK lawnmower industry can vary, but starts at around £1 billion. In 20 years, the over-Ten fishing fleet's fishing effort has decreased by 34 percent; lawnmower sales are set to increase by 7 percent in the next five years. And the fishing industry is subsidized. Marine Management Organization (UK Sea Fisheries Statistics, 2023), p.6.
11. 'Preventing Seabird Bycatch Near Bempton Cliffs', Mindfully Wired Communications, 25 August 2015, 3.34 minutes, https://www.youtube.com/watch?v=wjPPB5Xm0Q4.
12. BirdLife International, 'Troubled Waters: Fishing Bycatch Is Decimating Europe's Seabirds', 20 January 2022.
13. David Kroodsma et al., 'Global Prevalence of Setting Longlines at Dawn Highlights Bycatch Risk for Threatened Albatross', *Biological Conservation* 283 (2023), 110026.
14. Daniel Pauly, 'An Antidote to High-Tech Fishing', *Save Our Seas*.

3. FAST FISH FOOD

1. George Orwell, *The Road to Wigan Pier* (William Collins, 2021), Kindle, loc 1186.
2. University of Oxford Faculty of History, 'Rationing in Britain during World War II: A Resource for Key Stage 4'.

3   John Walton, *Fish and Chips and the British Working Class, 1870–1940* (Leicester University Press, 1992), p.86.
4   Ibid., p.87.
5   Ibid., p.77.
6   National Federation of Fish Friers, 'The Guide to Quality Fish and Chip Shops', 2021.
7   National Federation of Fish Friers, 'Everything You Need to Know About Fish and Chips'.
8   'Fish and Chips', *Daily Herald*, 17 January 1927.
9   British Chamber of Commerce in Japan, 'Welcome: Malins', 27 February 2023.
10   James Alexander, 'The Unlikely Origin of Fish and Chips', *BBC News*, 16 December 2009.
11   'Fish and Chips', *Aberdeen Press and Journal*, 2 November 1927.
12   Barton Seaver, *American Seafood: Heritage, Culture & Cookery from Sea to Shining Sea* (Sterling Epicure, 2017), p.23.
13   National Fisheries Institute, 'NFI Annual Top 10 List Illustrates Record Year for Seafood Consumption in 2021', 7 June 2023.
14   William Bradford, *Bradford's History of the Plymouth Settlement 1608–1650* (E. P. Dutton, 1909), p.4.
15   F. W. Bell, 'The Pope and the Price of Fish', *The American Economic Review* 58, no. 5 (December 1968), 1346–50.
16   Ute Eberle, 'The Unlikely Success of Fish Sticks', *Hakai Magazine*, 23 April 2021.
17   Paul Josephson, 'The Ocean's Hot Dog: The Development of the Fish Stick', *Technology and Culture*, 49, no. 1 (January 2008), 41–61.
18   'Fish Fingers', *Rigby's Encyclopaedia of the Herring*.
19   'Inside Europe's Largest Fish Finger Factory', posted by Free Doc Bites, 13 August 2021, 12.40 minutes, https://www.youtube.com/watch?v=jNEFsryes4E.
20   Felicity Cloake, 'Fish Fingers Turn 60: How Britain Fell for Not-Very-Fishy Sticks of Frozen Protein', *Guardian*, 15 September 2015.
21   Judith Burns, '"Cheese Is from Plants" – Study Reveals Child Confusion', *BBC News*, 3 June 2013.
22   'One Fifth of Young Adults Think Fish Fingers ACTUALLY ARE the Fingers of Fish, Research Finds', *Daily Mirror*, 22 November 2015. This was filed under 'Weird News'.

Notes

23 Birds Eye, '20 Omega 3 Fish Fingers'.
24 Cloake, 'Fish Fingers Turn 60'.
25 Eberle, 'Unlikely Success of Fish Sticks'.
26 Catherine Feverherd, 'Everybody's Raving About Birds Eye Fish Sticks – The Only Fish Sticks of Sweet Ocean Perch!', *Kansas City Times*, 11 February 1955.
27 'Fish Fingers', *Rigby's Encyclopaedia*.
28 Gortons, 'Our Story'.
29 Kara Baskin, 'The Fish Stick Is Going Modern to Fit Pandemic Tastes', *Boston Globe*, 25 August 2020.
30 Janelle Nanos, 'The Gorton's Fisherman Has a New Crew', *Boston Globe*, 6 March 2019.
31 Walton, *Fish and Chips*, p.89.
32 The offensive trades legislation was designed to prevent the establishment of offensive or nuisance trades and to monitor existing establishments. Under the Public Health Act of 1936, other offensive trades included blood boiling and blood drying, bone boiling, fat extraction, glue making, gut scraping, soap boiling, and tallow melters.
33 'Fish and Chips', *Birmingham Gazette*, 13 December 1927.
34 Walton, *Fish and Chips*, p.149.
35 'Fish and Chips', *Hull Daily Mail*, 5 March 1936.
36 Walton, *Fish and Chips*, p.150.
37 Ibid., p.7.
38 Philip Howard, 'A Century of Fish and Chips', *The Times*, 2 September 1968.
39 Jennifer Jacquet and Daniel Pauly, 'Reimagining Sustainable Fisheries', *PLoS Biology*, 20, no. 10 (2022), e30018291829.
40 Suzi Pegg-Darlison, 'Fish and Chips: Two-Year Overview Ending Sept. '22', *Seafish*, 31 October 2022.
41 UK Geographics, 'Social Grade A, B, C1, C2, D, E', 23 February 2014.
42 Pegg-Darlison, 'Fish and Chips'.
43 Stephen White, 'End in Sight for British Chippies As HALF Are Expected to Close in Three Years', *Mirror*, 2 June 2022.
44 'Chip Shops Face "Extinction" Amidst Cost of Living Crisis', *BBC News*, 24 August 2022.

4. WE'RE ROLLING OVER

1. National Transportation Safety Board, 'Marine Board of Investigation into the Sinking of the *Scandies Rose* on December 31, 2019', 24 February 2021, p.548.
2. US Coast Guard, 'Report of the Marine Board of Investigation into the Commercial Fishing Vessel *SCANDIES ROSE* (O.N. 602351) Sinking and Loss of the Vessel with Five Crewmembers Missing and Presumed Deceased South of Sutwik Island, Alaska on December 31, 2019', MISLE Activity Number: 6881487, 1/11/2023, p.x.
3. 'Crab Boat Survivor Shares Story', posted by Gabe Cohen, 16 January 2020, 2.59 minutes, https://www.youtube.com/watch?v=CbKRj1bQLBo.
4. Unless otherwise specified, Dean Gribble's narrative is taken from his witness testimony to the National Transportation Safety Board investigation into the sinking of F/V *Scandies Rose*, accident no. DCA20FM009, 1 January 2020.
5. Centers for Disease Control and Prevention (CDC)/National Institute for Occupational Safety and Health (NIOSH), 'Commercial Fishing Safety: Alaska', 4 December 2023.
6. Royal Meteorological Society, 'The Beaufort Wind Scale'.
7. CDC/NIOSH, 'About Commercial Fishing Safety', 13 December 2023.
8. Marine Accident Investigation Branch, 'Safety Digest: Lessons from Marine Accident Reports', 1/2024, p.15.
9. Ibid., p.12.
10. US Congress, Senate and House, 'Commercial Fishing Industry Vessel Safety of 1988', 102 Stat. 1585, Public Law 100-424, approved 9 September 1988.
11. Maritime and Coastguard Agency, 'Fishing Certification and Training', 19 September 2012, updated 20 April 2021.
12. Code of Federal Regulations, 46 CFR 141.350.
13. National Transportation Safety Board, 'Improve Fishing Vessel Safety', 14 October 2021.
14. Riley Woodford, 'Fatality-Free Year for Alaska Fishing', *Alaska Fish & Wildlife News*, November 2022.
15. Royal National Lifeboat Institution, 'Choose It, Wear It', p.2.
16. John J. Poggie and Richard B. Pollnac, 'Safety Training and Oceanic Fishing', *Marine Fisheries Review*, 59, no. 2 (1997).

Notes

17 Zaz Hollander and Aubrey Wieber, 'From Sleeping to Swimming, It Was About 10 Minutes, Says Survivor of Fishing Boat Sinking', *Anchorage Daily News*, 3 January 2020.

18 National Transportation Safety Board, 'Interview of Gerry Cobban Knagin, Sister of *Scandies Rose* Captain', 5 January 2020.

19 Anna Grybenyuk, 'Deadly Work: The Dangers of Fishing and How to Make It Safer', Lloyds Register Foundation Heritage and Education Centre, 14 November 2023.

20 Food and Agriculture Organization of the United Nations, 'About Fishing Safety'.

21 Fish Safety Foundation, 'Triggering Death: Quantifying the True Cost of Global Fishing', White Paper, November 2022.

22 International Convention for the Safety of Life at Sea, 1974, as amended, p.284.

23 International Maritime Organization, '2012 Cape Town Agreement to Enhance Fishing Safety'.

24 Maritime & Coastguard Agency, 'Fisherman's Safety Guide' (6 May 2014), p.9.

25 'Size Counts', *Fishermen's News*, 1 April 2022.

26 'Donning an Immersion Suit', posted by US Coast Guard, 2 November 2018, 2.09 minutes, https://www.youtube.com/watch?v=doRLksdmYCE.

27 Theodore Teske, Samantha Case, Devin Lucas, Christy Forrester, and Jennifer Lincoln, 'Have You Met Angus? Development and Evaluation of a Social Marketing Intervention to Improve Personal Flotation Device Use in Commercial Fishing', *Journal of Safety Research*, 83 (2022), 260–8.

28 'Pacific Marine Expo 2019 Fishermen of the Year Contest', posted by Alaska Marine Safety Education Association, 14 December 2019, 2.26 minutes, https://www.youtube.com/watch?v=LTxNeDjJ4mI.

29 CDC Crew Member Survival Info Sheet, 'Increasing Your Chances of Surviving a Vessel Sinking'.

30 Docket 78, Scandies Rose Ch 13 MAYDAY 2020-01-01 0655GMT.

31 Frank Golden and Michael Tipton, *Essentials of Sea Survival* (Human Kinetics, 2002), p.63.

32 Northeast Center for Occupational Health and Safety, 'Lifejackets for Lobstermen: Summary Report 2020'.

33 'Fatal Falls Overboard in Commercial Fishing, United States, 2000–2016', *CDC Morbidity and Mortality Weekly*, 67, no. 16 (27 April 2018), 465–9.

34 In fact, this quote is judged by the Churchill Project to be 'undocumented but not uncharacteristic'. https://winstonchurchill.hillsdale.edu/americans-will-always-right-thing/.

35 US Coast Guard, 'Report of the Marine Board of Investigation into the Commercial Fishing Vessel *SCANDIES ROSE*', p.29.

36 Ibid., p.30.

37 '*Scandies Rose*: Coast Guard Hearings', *National Fisherman*, 18 February 2021.

38 US Coast Guard, 'Operational Safety Information for Commercial Crabbing Vessels 2024'.

39 Hal Berton, 'Settlement in Deadly Sinking of *Scandies Rose* Crab Boat Calls for More Than $9 Million Payout', *Seattle Times*, 2 November 2020.

40 Hope McKenney, 'Man Who Survived the Sinking of the *Scandies Rose* Dies in Anchorage Motorcycle Crash', KUCB, 3 November 2021.

41 https://www.gofundme.com/f/jon-lawler-memorial-fund.

5. A COARSE AND VULGAR WOMAN

1 Sébastien-Roch-Nicolas Chamfort, *Chamfort: Reflections on Life, Love & Society: Together with Anecdotes and Little Philosophical Dialogues* (Short Books, 2003), p.134.

2 Food and Agriculture Organization of the United Nations, 'The Role of Women in the Seafood Industry' (FAO, 2015), p.8.

3 Marie Christine Monfort, 'Let's Acknowledge Invisible, Ignored and Unrecognised (IIU) Women in the Seafood Industry'.

4 Liam Thorp, 'Mayor Joe Anderson Reported for Alleged "Sexist" Remark', *Liverpool Echo*, 16 November 2017.

5 *Oxford English Dictionary*, virago.

6 Henry Mayhew, *London Labour and the London Poor*, vol. 1 (Charles Griffin, 1864; Project Gutenberg, 2017), p.11.

7 'It is ordered that no *birlester* who carries oysters, mussels, salt fish, and other victuals, in the City to sell, shall stand in any street or lane of the said city, nor yet in his shop, to retail them; but such person shall be always moving about in the said city from street to street, and from lane to lane, to retail the same; on pain of forfeiting all the

victuals found on sale as against this Ordinance', 'Regulations as to the Sale of Fish in the City; and as to the Use of Nets in the Thames', 12 Richard II. A.D. 1388. Letter-Book H. fol. Ccxxxvi, *Memorials of London and London Life in the 13th, 14th and 15th Centuries* (Longmans, Green, London, 1868). A 'birlester', actually any kind of peddler, was used to describe women who sold fish.

8 Christi Spain-Savage, 'The Gendered Place Narratives of Billingsgate Fishwives', *SEL Studies in English Literature 1500–1900*, 56, no. 2, (Spring 2016), 417–34.

9 Carol Midgley, 'An Audience with Adele Review – Cor Blimey, Our Favourite Girl Next Door Still Charms', *The Times*, 21 November 2021.

10 Walter Scott, *The Antiquary* (Archibald Constable, 1816; Project Gutenberg, 2016), chapter twenty-six.

11 Margaret King, 'A Partnership of Equals: Women in Scottish East Coast Fishing Communities', *Folk Life*, 31, no. 1 (1992), 17–35.

12 John Gray Centre, 'Fishwives of East Lothian'.

13 'Guinness World Records' [London]: *Guinness World Records*, 11 February 2024.

14 Sir John Sinclair, *The Statistical Account of Scotland, Inveresk, Edinburgh*, vol. 16 (William Creech, 1795), p.20.

15 Sinclair, *Statistical Account of Scotland*, p.19.

16 King, 'A Partnership of Equals', 20.

17 Sunniside Local History Society, 'The Cullercoats Fishwife'.

18 Margaret Bochel, *Nairn Speldings: The Story of Nairn Fishwives* (Nairn Fishertown Museum, 1998), p.17.

19 King, 'A Partnership of Equals'.

20 'Fisherwomen's Trip', *Musselburgh News*, 7 July 1939. In the next column, 'The Inveresk Duckling Mystery Has Been Solved'.

21 Marine Stewardship Council, 'What Is Herring?'

22 Both 608 Squadron (based in North Yorkshire) and 206 Squadron (based in Bircham Newton in East Anglia), were nicknamed Kipper Patrols. See UK Government, Ministry of Agriculture and Fisheries, 'Fisheries in War Time: Report on the Sea Fisheries of England and Wales for the Years 1939–1944 inclusive', (HMSO, 1946), p.33.

23 Georg Pfeifer, 'Störung des Mietgebrauchs durch Mieter', 15 February 2005.

24　Gordon Johnston, 'Gyaan sooth tae da' guttin'', *Unkans*, Shetland Museum and Archives and the Shetland Heritage Community, no. 32 (May 2012), 4.
25　Johnston, 'Gyaan sooth tae da' guttin''.
26　Buckie and District Fishing Heritage Centre, 'Interviews: Maggie Cowie'.
27　Buckie and District Fishing Heritage Centre, 'Interviews: Jeannie Innes'.
28　King, 'A Partnership of Equals'.
29　Buckie and District Fishing Heritage Centre, 'Interviews: Nanie Kaeczmarek'.
30　King, 'A Partnership of Equals'.
31　James E. Henderson Ltd, producer, 'North Sea Herring Fleet', 1935, 6.53 minutes, retrieved from https://movingimage.nls.uk/film/3110.
32　Thompson, 'Women in the Fishing'.
33　Steve Humphries, director, 'Hull's Headscarf Heroes', BBC Four (2018), 59 minutes.
34　UK Government, 'Fishing Vessels (Safety Provisions) Bill', vol. 797, cc276 (3 March 1970), p.324.
35　Humphries, 'Hull's Headscarf Heroes'.
36　Fishing Vessels (Safety Provisions) Bill.
37　Trevor Gibbons, 'The Triple Trawler Tragedy: The Hull Fishermen Who Never Came Home', *BBC News*, 4 February 2018.
38　'Sobbing Wife Brings a Trawler Victory', *People*, 4 February 1968.
39　Brian W. Lavery, *The Headscarf Revolutionaries: Lillian Bilocca and the Hull Triple-Trawler Disaster* (Barbican Press, 2024), p.104.
40　Humphries, 'Hull's Headscarf Heroes'.
41　Ibid.
42　'Sobbing Wife Brings a Trawler Victory'.
43　Lavery, *The Headscarf Revolutionaries*, p.119.
44　John Chartres, 'Clash Expected at Trawler Talks', *The Times*, 8 February 1968.
45　'Sobbing Wife Brings a Trawler Victory'.
46　John Young, 'Trawlermen's Wives to Try "Diplomacy"', *The Times*, 3 February 1968.
47　Lavery, *The Headscarf Revolutionaries*, p.170.

Notes

48 Ibid., p.170.
49 Humphries, 'Hull's Headscarf Heroes'.
50 'We're Going: Give Them All Our Love', *Daily Mirror*, 6 February 1968.
51 'Lone Survivor Relieves Arctic Night Ordeal', *Western Daily Press*, 16 October 1968.
52 Lavery, *The Headscarf Revolutionaries*, p.132.
53 'Mrs Bilocca Loses Her Job', *The Times*, 20 February 1968.
54 Marjorie Proops, 'The Real Big Lil', *Daily Mirror*, 8 February 1968.
55 Frank Scott, 'The Headscarf Revolutionaries: Lillian Bilocca and the Hull Triple-Trawler Disaster', *The Mariner's Mirror*, 103, no. 2 (December 2020), 251–2.
56 HM Government, Parliament Debate, 'Trawler "Kingston Peridot"', *Hansard*, HC Deb 31, vol. 757, cc1347–50 (January 1968).
57 'Royal Commission on Trawling Demanded', *Hull Daily Mail*, 1 February 1968.
58 Brian Lavery, 'Triple Trawler Disaster: Hull's Headscarf Revolutionaries', *Fishing News*, 7 October 2022.
59 Lavery, *The Headscarf Revolutionaries*, p.212.
60 '"Big Lil" Bilocca Dies in Hospital', *Hull Daily Mail*, 5 August 1988.
61 Dan Kemp, 'William Wilberforce Is Named as the Greatest Hullensian Ever in Poll', *Hull Daily Mail*, 8 March 2021.
62 Humphries, 'Hull's Headscarf Heroes'.

6. TRAWLING

1 H. Bruce Franklin, *The Most Important Fish in the Sea: Menhaden and America* (Island Press, 2007), p.127.
2 European Union, IUU position paper, 'The Need for Mandatory IMO Numbers for Vessels Catching Seafood for the EU Market', May 2017.
3 Marine Conservation Institute, 'Are Fish Aggregating Devices Just a FAD?', 29 July 2021.
4 Blue Marine Foundation, 'More Than 100 NGOs Call for Improved Management of Fish Aggregating Devices', 4 October 2021.
5 Charles Dickens, *On Travel* (Hesperus Press, 2009), p.19.
6 Doreen Huppert, Judy Benson, and Thomas Brandt, 'A Historical View of Motion Sickness – A Plague at Sea and on Land, also with Military Impact', *Frontiers in Neurology* 8 (2017), 114.

7   Cary Balaban, Sara Ogburn, Susan Warshafsky, Abdul Ahmed, and Bill Yates, 'Identification of Neural Networks that Contribute to Motion Sickness through Principal Components Analysis of Fos Labeling Induced by Galvanic Vestibular Stimulation', *PLoS One*, 9, no. 1 (23 January 2014), e86730.

8   Allan Hall, 'Fish Get Seasick, Scientist "Proves"', *Telegraph*, 20 April 2009.

9   Eduardo Santurtun, Grisel Navarro, and Clive Phillips, 'Do Antiemetics Attenuate the Behavioural Responses of Sheep to Simulated Ship Motion?', *Applied Animal Behaviour Science*, 223 (2020), 104924.

10   Mary Roach, *Packing for Mars: The Curious Science of Life in the Void* (W. W. Norton, 2010), p.125.

11   Hall, 'Fish Get Seasick, Scientist "Proves"'.

12   International Council for the Exploration of the Sea, 'Roadmap for ICES Bycatch Advice on Protected, Endangered and Threatened Species', 2022.

13   Grant Course, Johanna Pierre, Belinda Howell, and the WWF, 'What's in the Net? Using Camera Technology to Monitor, and Support Mitigation of, Wildlife Bycatch in Fisheries', 2020.

14   Stella Nemecky, 'The Untrawled Truth: Why EU Fisheries (Control) Policy Should Strengthen Discard Monitoring, Control and Reporting within an Implemented Landing Obligation', WWF Germany, 2022.

15   Qianqian Tao, 'Swim Bladder: Would Being Stabbed be Beneficial to a Fish?', *SQ*, 5 March 2023.

16   Tim Martindale, 'Livelihoods, Craft and Heritage: Transmissions of Knowledge in Cornish Fishing Villages' (PhD dissertation, University of London, 2012), 174.

17   Richard Mackay, Christopher McEntyre, Caroline Henderson, Michael Lever, and Peter George, 'Trimethylaminuria: Causes and Diagnosis of a Socially Distressing Condition', *The Clinical Biochemical Reviews*, 32, no. 1 (February 2011), 33–43.

18   News Dog Media, 'Fish Odour Syndrome: Woman Suffers Rare Condition That Makes Her Smell Like Rotting Fish', *HuffPost*, 3 July 2014.

19   Cornwall Wildlife Trust/Cornwall Good Seafood Guide, 'Lesser Spotted Dogfish'.

20   Seafood Cornwall, 'The Great British Seafood Dilemma', May 2022.

21   William Warner, *Distant Water: The Fate of the North Atlantic Fisherman* (Little, Brown, 1983).

7. FISH ARE NOT CHIPS

1   Richard Fidler, host, 'Dr Fish Feelings', Australian Broadcasting Corporation, 18 January 2021, 51 minutes.

2   Redouan Bshary, 'Machiavellian Intelligence in Fishes', in *Fish Cognition and Behaviour*, 2nd edition, ed. Culum Brown, Kevin Laland, and Jens Krause (Wiley-Blackwell, 2011).

3   Redouan Bshary and Culum Brown, 'Fish cognition', Current Biology, 24, no. 19 (2014), R947–50.

4   Masanori Kohda, Takashi Hotta, Tomohiro Takeyama et al., 'Cleaner Wrasse Pass the Mark Test. What Are the Implications for Consciousness and Self-Awareness Testing in Animals?', *PLOS Biology* (7 February 2019).

5   'This Fish Knows Its Own Face in a Mirror', *Nature* 614, no. 393 (2023), 2.

6   Masanori Kohda, Redouan Bashary, Naoki Kubo, and Satoshi Awata, 'Cleaner Fish Recognise Self in Mirror via Self-Face Recognition Like Humans', *Proceedings of the National Academy of Sciences*, 120, no. 7 (6 February 2023).

7   Culum Brown, 'Fish Intelligence, Sentience and Ethics', *Animal Cognition*, 18, no. 1 (January 2015), 1–17.

8   Brown, 'Fish Intelligence'.

9   University of Sydney, '13th Robert Dixon Animal Welfare Symposium', 14 September 2023, 1:28.

10   Lewis & Clark Law School, 'World Aquatic Animal Day', 2022.

11   20,000 tonnes – no individuals of course – are now caught on average each year. Mark Cawardine, 'The World's First Commercial Octopus Farm Is a Disaster Waiting to Happen', *BBC Wildlife*, 23 March 2022.

12   Brown, 'Fish Intelligence'.

13   Robert Dixon Animal Welfare Memorial Symposia.

14   Ren Ryba and Sean Connell, 'Animal Minds, Social Change, and the Future of Fisheries Science', *Frontiers in Marine Science*, 8 (2021).

15   Stuart Harratt, 'Eating Fish Same as Eating Cats, Cleethorpes Animal Rights Poster Suggests', *BBC News*, 26 April 2023.

16   Chelsea Munro, 'PETA Billboard to Grimsby Locals: Eating a Fish Is Like Eating a Cat', PETA UK, 23 August 2023.

17 Harratt, 'Eating Fish Same as Eating Cats'.
18 Peter Singer, *Animal Liberation* (Random House, 1995), p.174.
19 Peter Singer, 'If Fish Could Scream', *Project Syndicate*, 13 September 2010.
20 William Strange, 'Can Fish Feel Pain?', *Gentleman's Magazine*, September 1870.
21 Brian Key, 'Why Fish Do Not Feel Pain', *Animal Sentience*, 1, no. 3 (2016).
22 Carl Safina, 'Are We Wrong to Assume Fish Can't Feel Pain?', *Guardian*, 30 October 2018.
23 Jonathan Balcombe, *What a Fish Knows: The Inner Lives of Our Underwater Cousins* (One World, 2017).
24 Sy Montgomery, 'Deep Intellect: Inside the Mind of the Octopus', *Orion*, November/December 2011.
25 V. Schluessel, N. Kreuter, I. M. Gosemann et al., 'Cichlids and Stingrays Can Add and Subtract "One" in the Number Space from One to Five', *Scientific Reports* 12, no. 3894 (2022).
26 Brown, 'Fish Intelligence', 9.
27 Shachar Givon, Matan Samina, Ohad Ben-Shahar, and Ronen Segev, 'From Fish out of Water to New Insights on Navigation Mechanisms in Animals', *Behavioural Brain Research*, 419 (2022), 113711.
28 Lily Carey, 'Goldfish May Have a Longer Memory Than Just Three Seconds', *Discover*, 2 April 2024.
29 Brown, 'Fish Intelligence', 8.
30 Government of Canada/Gouvernement du Canada, 'Wolffish: What You Need to Know'.
31 Callum Roberts, 'Britain's Real Fish Fight', *Guardian*, 13 January 2011.
32 Tom Murray, 'We're Accidentally Driving This Extremely Ugly Fish to Extinction', *Business Insider*, 14 January 2019.
33 Municipality of Bodø, Nordland, 'Regulation on the Protection of Saltstraumen Marine Conservation Area, Bodø', Entry into Force 21 June 2013.

8. A TIN-CAN NAVY

1 'Scalloper Crew Save 31 Lives in Channel Rescue', *Fishing News*, 20 December 2022.

Notes

2 UK Government, 'Monthly Number of Migrants Detected in Small Boats – 1 December to 31 December 2022'.

3 Diane Taylor and Dan Sabbagh, '"You Could See the Panic": How the Channel Small Boat Incident Unfolded', *Guardian*, 14 December 2022.

4 Matt Dathan, Adam Sage, and Charlotte Wace, '"Please Help. We Have Children in a Boat and Water Is Coming"', *The Times*, 15 December 2022.

5 Ben Borland, 'Fuming Scots Fishing Boat Skipper Who Rescued Migrants Says UK Authorities Acting As "Taxi Service"', *Scottish Daily Express*, 19 December 2022.

6 The Refugee Council, 'Understanding Channel Crossings'.

7 Royal National Lifeboat Institution, 'RNLI Statement on Channel Incident', 15 December 2022.

8 Robb Robinson, *Trawling: The Rise and Fall of the British Trawl Fishery* (University of Exeter Press, 1998), pp.118–19.

9 Robb Robinson, *Fishermen, the Fishing Industry and the Great War at Sea* (University of Liverpool Press, 2019), p.16.

10 Robinson, *Fishermen*.

11 J. W. B. Chapman, 'The Lighter Side of Patrol Duties', *Daily Mail*, 16 September 1916.

12 Harry Ludlam and Paul Lund, *I Was There: When the Trawlers Went to War* (W. Foulsham, 1978), Kindle, loc 513.

13 Scarborough Maritime Heritage Centre, 'German U-boat Sinks 11 Scarborough Trawlers in One Night – World War One'.

14 Charles Dana Gibson, 'Victim or Participant? Allied Fishing Fleets and U-boat Attacks in World Wars I and II', *The Northern Mariner/Le Marin du nord*, I, No. 4 (October 1991), 1–18.

15 Tom Tulloch-Marshall, 'Q-ships'.

16 Robb Robinson, *Fishermen, the Fishing Industry and the Great War at Sea* (University of Liverpool Press, 2019), p.179.

17 Ludlam and Lund, *I Was There*, loc 157, 196, 243.

18 Ibid., *I Was There*, loc 1661.

19 Joe Steele, oral interview, Imperial War Museum Reel 4.

20 Harry Tate's Navy, 'The Royal Naval Patrol Service at War'.

21 Georgina Rayner, 'Harry Tate's Navy', *BBC*, 12 December 2005.

22 'Trawler's Capture of U-boat, the Captain's Story', *The Times*, 11 October 1941.

23 Commonwealth War Graves Commission, 'Lowestoft Naval Memorial'.
24 'Lowestoft Naval Memorial'.

9. BEFORE, THERE WAS FISH

1 Food and Agriculture Organization of the United Nations, 'Decent Rural Employment: Senegal'.
2 Missing Migrants Project, 'Migration with Africa'.
3 Julia Black, 'Maritime Migration to Europe: Focus on the Overseas Route to the Canary Islands' (International Organization for Migration, 2021).
4 'In Senegal, a Village Mourns Its Sons Dead at Sea in Migrant Tragedy', *France 24*, 17 August 2023.
5 Wahany Johnson Sambou, 'Senegal: le village de Fass Boye pleure ses morts en mer', *AfricaNews*, 13 August 2023 [my translation from French].
6 Grace Ekpu and Patrick Whittle, 'Senegal Struggles with Loss of Fish Central to Diet, Culture', *Associated Press*, 6 April 2024.
7 Djiga Thiao, Christian Chaboud, Alassane Samba, Francis Laloë, and Phillipe Cury, 'Economic Dimension of the Collapse of the "False Cod" *Epinephelus aeneus* in a Context of Ineffective Management of the Small-Scale Fisheries in Senegal', *African Journal of Marine Science*, 34, no. 3 (2012), 305–11.
8 International Union for the Conservation of Nature Shark Specialist Group, 'New Global Study Finds Unprecedented Shark and Ray Extinction Risk', 6 September 2021.
9 'Unprecedented Shark and Ray Extinction Risk'.
10 Food and Agriculture Organization of the United Nations, 'Coastal Fisheries Initiative: CFI in Senegal'.
11 El hadj Bara Deme, Moustapha Deme, and Pierre Failler, 'Small Pelagic Fish in Senegal: A Multi-usage Resource', *Marine Policy*, 141 (2022), 105083.
12 Joelle Philippe, 'Senegal's Exports of Fishmeal and Fish Oil "Explode"', Coalition for Fair Fisheries Arrangements, 13 June 2024.
13 European Parliament, 'Protocol on the Implementation of a Sustainable Fisheries Partnership between the European Union and the Republic of Senegal', 20 November 2019.

Notes

14   Greenpeace, 'Europe Wants Senegal's Fish but Rejects Its Migrants', 19 November 2020.

15   European Parliament, 'Implementation of Sustainable Fisheries Partnership'.

16   Dyhia Belhabib, U. Rashad Sumaila, Vicky Lam et al., 'Euros vs. Yuan: Comparing European and Chinese Fishing Access in West Africa', *PLOS ONE*, 10, no. 3 (2015).

17   Abdoulaye Sarre, Hervé Demarcq, Noel Keenlyside et al., 'Climate Change Impacts on Small Pelagic Fish Distribution in Northwest Africa: Trends, Shifts, and Risk for Food Security', *Scientific Reports*, 14 (2024), 12684.

18   Environmental Justice Foundation, 'New Trawlers with History of Illegal Fishing Threaten Senegalese Fisheries', 22 April 2020.

19   Food and Agriculture Organization of the United Nations, Fishery Committee for the Eastern Atlantic, 'Summary Report. FAO Working Group on the Assessment of Small Pelagic Fish Off Northwest Africa 2019', p.9.

20   Greenpeace Africa, 'Procès historique contre l'industrie de la farine et de l'huile de poisson: une usine pollue l'environnement et déverse illégalement des déchets dans un lac', 7 October 2022.

21   Jack Thompson, 'The Fervent Fight Over Fish Meal', *Hakai Magazine*, 11 August 2023.

22   Fouad Reda, Hilmar Kjartansson, and Steven Jeffery, 'Use of Fish Skin Graft in Management of Combat Injuries Following Military Drone Assaults in Field-Like Hospital Conditions', *Military Medicine*, 188 (3 November 2023), 11–12.

23   Zara Rubin, 'Waitress, 36, Who Suffered Severe Burns While She Was Working Has Her Wounds Dressed with FISH SKIN in a Pioneering New Treatment', *Daily Mail*, 16 December 2016.

24   University of California at Davis, 'Dogs, Cats, Rescued from California Camp Fire with Fish Skins', 12 December 2018.

25   Reda, 'Use of Fish Skin Graft'.

26   Food and Agriculture Organization of the United Nations, Fisheries Department, 'Aquaculture Development. 1. Good Aquaculture Feed Manufacturing Practice', FAO Technical Guidelines for Responsible Fisheries, no. 5, suppl. 1 (2001).

27   American Association of Feed Control Officials, *Official Publication: Pet Food* (AAFCO, 2022), Chapter 6.

28 Changing Markets Foundation/Greenpeace Africa, 'Feeding a Monster: How European Aquaculture and Animal Feed Industries Are Stealing Food from West African Communities', June 2021, p.5.

29 Changing Markets Foundation/Compassion in World Farming, 'Until the Seas Run Dry', April 2019, p.7.

30 Tim Cashion, Frédéric Le Manach, Dirk Zeller, and Daniel Pauly, 'Most Fish Destined for Fishmeal Production Are Food-Grade Fish', *Fish and Fisheries*, 18 (2017), 837–44.

31 OECD/FAO, *OECD-FAO Agricultural Outlook 2021-2030* (OECD Publishing, 2021).

32 Changing Markets Foundation, 'Fishing for Catastrophe: How Global Aquaculture Supply Chains Are Leading to the Destruction of Wild Fish Stocks and Depriving People of Food in India, Vietnam and The Gambia' (2019), p.5.

33 Council Regulation (EC) No 1005/2008 of 29 September 2008.

34 'Fishing for Catastrophe', p.4.

35 The Marine Ingredients Organisation, 'The Role of Marine Ingredients'.

10. DOWN BELOW

1 NatureScot/NàdarAlba, 'Maerl Beds'.

2 NatureScot/NàdarAlba, 'Priority Marine Features in Scotland's Seas – The List', 2020.

3 NatureScot/NàdarAlba, 'Marine Protected Areas'.

4 Damian Carrington, 'Marine Life Worse Off Inside "Protected" Areas, Analysis Reveals', *Guardian*, 20 December 2018.

5 Oceana, 'All But Two of Scotland's Offshore Marine "Protected" Areas Are Paper Parks', 4 August 2021.

6 Oceana, 'All But Two of Scotland's Offshore Marine "Protected' Areas Are Paper Parks', press release, 8 April 2021.

7 Veronica Relano, Tiffany Mak, Shelumiel Ortiz, and Daniel Pauly, 'Stakeholder Perceptions Can Distinguish "Paper Parks" from Marine Protected Areas', *Sustainability* 14 (2022), 9655.

8 Colin Rowat, personal communication.

9 Seafish, 'Fishing Data and Insight'.

10 Leigh Michael Howarth and Bryce Donald Stewart, 'The Dredge Fishery for Scallops in the United Kingdom (UK): Effects on Marine Ecosystems and Proposals for Future Management', Report to the

	Sustainable Inshore Fisheries Trust, Marine Ecosystem Management Report no. 5, University of York, May 2014, p.3.
11	Howarth and Stewart, 'Dredge Fishery for Scallops', p.11.
12	Michael Kaiser, K. Robert Clarke, Hilmar Hinz et al., 'Global Analysis and Recovery of Benthic Biota to Fishing', *Marine Ecology Progress Series*, 311 (2006), 1–14.
13	Aryn Baker, 'How Industrial Fishing Creates More CO_2 Emissions Than Air Travel', *Time*, 17 March 2021.
14	UK Government, 'Inshore Fishing (Scotland) Act 1984', Chapter 26.
15	Open Seas, 'Dredging in the Dark', 12 February 2019.
16	Government of Scotland, 'Freedom of Information Release, MPA Fisheries Management Rules: EIR Release', 17 May 2018.
17	Christopher Sleight, 'Rare Loch Carron Flame Shell Reef "Devastated" by Scallop Dredger', *BBC News*, 24 April 2017.
18	NatureScot/NàdarAlba, 'Flame-Shell Beds'.
19	Danielle Sloan, Catherine Jones, Leslie Noble, and Berit Rabe, 'Flame Shell Beds: A Protected Keystone Biogenic Bivalve Habitat', *Imperiled: The Encyclopedia of Conservation*, 1–3 (2022), 693–9.
20	Open Seas, 'A Lesson from Loch Carron', 27 May 2019.
21	Billy Briggs, 'Scotland's Largest Fishing Organisation Refuses to Clarify Role of Illegal Dredger', *Ferret*, 19 May 2021.
22	Daniel Pauly, 'Aquacalypse Now', *New Republic*, 28 September 2009.
23	Scottish Creel Fishermen's Federation, '3 Mile Limit – A Case for a Sustainable Fishery'.
24	Marine Conservation Institute, Marine Protection Atlas, https://mpatlas.org/

11. A DEDICATED FISH WARDEN

1	Human Rights at Sea, 'Fisheries Observer Deaths at Sea, Human Rights and the Role and Responsibilities of Fisheries Organisations', 1 July 2020.
2	Bernadette Carreon, 'Death at Sea: The Fisheries Inspectors Who Never Came Home', *Guardian*, 19 June 2021.
3	Human Rights at Sea, 'Independent Case Review into the Investigation of the Death of Kiribati Fisheries Observer Eritara Aati Kaierua', 19 May 2021.
4	Pew, 'Issue Brief: How to End Illegal Fishing', 12 December 2017.

5   Sarah Tory, 'Searching for Keith', *Hakai Magazine*, 9 July 2019.

6   Association for Professional Observers, 'Observer Deaths and Disappearances'.

7   Karen McVeigh, 'Disappearances, Danger and Death: What Is Happening to Fishery Observers?', *Guardian*, 22 May 2020.

8   Keith Granger Davis, Glenn David Quelch, and Anik Clemens, *Eyes on the Seas: A Look into the Fisheries Observer Profession Through Stories and Creative Works* (independently published), Kindle, 23.

9   Greenpeace US, 'Fisheries Observers Are Human Rights Defenders on the World's Oceans', briefing note, 25 November 2020.

10 Cu62329, 'Fisheries observer', *Reddit*, 12 September 2022, retrieved from https://www.reddit.com/r/marinebiology/comments/xcmtv8/comment/io617hd/.

11 Davis, Quelch, and Clemens, *Eyes on the Sea*, 77.

12 Davis, Quelch, and Clemens, *Eyes on the Sea*, 80.

13 Blake Fletcher, host, 'Half Hour Intern: Commercial Fishing Inspector', Episode 166, 1.01.34.

14 Jimmy Thomson, '"Trapped": Women Working as Fishery Observers Allege Sex Harassment, Assault at Sea', *Vice*, 9 February 2021.

15 Head_Adhesiveness879, 'TIL of fisheries observers who are employed to monitor fishing boat practices and compliance. They work alone on boats often thousands of kilometres from port. It is a dangerous job and they disappear often', *Reddit*, 27 December 2022, retrieved from https://www.reddit.com/r/todayilearned/comments/zwkmb6/til_of_fisheries_observers_who_are_employed_to/.

16 Unknown user, 'Is working as an at-sea fisheries observer dangerous?', *Reddit*, 4 August 2021, retrieved from https://www.reddit.com/r/marinebiology/comments/oxuzgn/is_working_as_an_atsea_fisheries_observer/.

17 Women in Ocean Science, 'Sexual Harassment in Marine Science', 8 March 2021, pp.3–4.

18 Renee Gross, 'NOAA Is Trying to Encourage More Observers to Report Sexual Harassment', *Alaska Public Media*, 5 June 2019.

19 Yuntao Wang and Jane diCosimo, National Observer Program, '2016 Fishery Observer Attitudes and Experiences Survey', National Oceanic and Atmospheric Administration Technical Memorandum NMFS-F/SPO-186, May 2019, p.3.

20 Jessica Dobson, Matthew Kaley, Anna Birkenbach, and Kimberly Oremus, 'Harassment and Obstruction of Observers in U.S. Fisheries', *Frontiers in Marine Science*, 13 September 2023.
21 Association for Professional Observers, 'Harassment'.
22 Alaska Fisheries Science Center and Alaska Regional Office, 'North Pacific Observer Program 2022 Annual Report', 2024, p.v.
23 Marine Stewardship Council, 'Review of Optimal Levels of Observer Coverage in Fishery Monitoring', May 2021, p.2.
24 Fletcher, Half Hour Intern Podcast.
25 Food and Agriculture Organization of the United Nations, 'Port State Measures'.
26 Food and Agriculture Organization of the United Nations Indian Ocean Tuna Commission, 'IOTC IUU List'.
27 Government of the Philippines, 'Tañon Strait'.
28 Allison Guy, 'Shot, Paralyzed, but Not Broken, a Filipino Activist Vows to Keep Fighting for the Ocean', Oceana Philippines, 18 July 2016.
29 Statista, 'Leading Fishery Exports from the Philippines in 2022, by Volume', 19 February 2024.
30 Oceana Philippines, 'Love Letter to the Tañon Strait', February 2015.
31 Amina Rasul, 'The Earth at Risk: "Bantay Dagat, Bantay Yaman"', *The Manila Times*, 27 April 2008.
32 Philippine Navy, 'Bantay Dagat (Sea Patrol) Forces'.
33 Antonio Oposa, 'Let Me Tell You a Story', *Daedalus*, 149, no. 4 (2020), 207–33.
34 Senate of the Philippines, 'A Tear and Justice for a Fallen Hero of the Forests', 16 May 2011.
35 Oceana Philippines, 'The Philippines Protects 266,000 Square Kilometers of Ocean from Bottom Trawling', 13 December 2018.
36 José Santino Bunachita, 'For the Love of the Sea', *Cebu Daily News*, 10 June 2016.

12. A VERY SLAVERY JOB ACTUALLY

1 Food and Agriculture Organization of the United Nations, *Code of Conduct for Responsible Fisheries* (FAO, 1995).
2 Antonia Cundy, 'The Fishermen', *Financial Times*, 15 June 2023.
3 International Labour Organization, *The Work in Fishing Convention, 2007 (No. 188)*.

4   International Transport Workers Federation, Briefing Paper, 'A One-Way Ticket to Labour Exploitation', May 2022.
5   Task Force on Human Trafficking in Fishing in International Waters, Report to Congress, January 2021.
6   Robin McDowell, Margie Mason, and Martha Mendoza, 'AP Investigation: Slaves May Have Caught the Fish You Bought', Associated Press, 25 March 2015.
7   Alistair Coupe, Hance Smith, and Bruno Ciceri, *Fishers and Plunderers: Theft, Slavery and Violence at Sea* (Pluto Press, 2015).
8   Environmental Justice Foundation, 'Blood and Water: Human Rights Abuse in the Global Seafood Industry', June 2019, p.7.
9   Sea Shepherd, '"They Ate Cats for Lunch": Harrowing Tales from One of Two Trawlers Arrested Off Liberian Coast', 31 October 2022.
10 US Government, Department of Labor, 'List of Goods Produced by Child Labor or Forced Labor'.
11 Ian Urbina, 'Subsidizing China's Fishing Fleet', *Outlaw Ocean Project*, 2 September 2021.
12 McDowell, Mason, and Mendoza, 'AP Investigation'.
13 Environmental Justice Foundation, 'Thailand's Seafood Slaves: Human Trafficking, Slavery and Murder in Kantang's Fishing Industry', 2015.
14 David Tickler, Jessica Meeuwig, Katharine Bryant et al., 'Modern Slavery and the Race to Fish', *Nature Communications*, 9, no. 1 (7 November 2018), 4643.
15 Marine Accident Investigation Branch, 'Annual Report 2022', p.55.
16 John Holden, 'Five "Slaves" Rescued from Fishing Boat in Shoreham Port', *Sussex World*, 5 November 2020.
17 Harry Bullmore, 'Slavery Arrests as Men Recovered from Fishing Boat in Shoreham', *Argus*, 5 November 2020.
18 Chris Clements, presenter. 'Slavery at Sea', *BBC*, 58 minutes, 19 August 2024, available from https://www.bbc.co.uk/programmes/m002271r.
19 Ibid.
20 Ibid.

13. A NICE BIT OF SALMON
1   Barents Watch, 'Localities' (undated), retrieved from https://www.barentswatch.no/fiskehelse/localities.
2   WWF, 'Tuna'.

Notes

3   Seafish, 'Atlantic Salmon'.

4   Melissa Aronson, 'Supporting Socially Responsible Farmed Salmon in Chile', 12 July 2024.

5   *Financial Times*, 'Equities: MOWI'.

6   Cécile Brugère and Neil Ridler, 'Global Aquaculture Outlook in the Next Decades: An Analysis of National Aquaculture Production Forecasts to 2030', FAO Fisheries Circular no. 1001, 2004, Chapter 2.

7   Food and Agriculture Organization of the United Nations, *The State of World Fisheries and Aquaculture 2024: Blue Transformation in Action* (FAO, 2024), p.121.

8   DSM-Firmenich, 'Carophyll® – Because Color Matters'.

9   Frode Alfnes, Atle Guttormsen, Gro Steine, and Kari Kolstad, 'Consumers' Willingness to Pay for the Color of Salmon: A Choice Experiment with Real Economic Incentives', *American Journal of Agricultural Economics*, 88 (2016), 1050–61.

10   Kurt Oddekalv, 'Report on the Environmental Impact of Farming of North Atlantic Salmon in Norway', Norges Miljøvernforbund (Green Warriors of Norway), 2011, p.6.

11   Chesapeake Bay Program, 'Lumpfish'.

12   Kathy Overton, Luke Barrett, Frode Oppedal, Tore Kristiansen, and Tim Dempster, 'Sea Lice Removal by Cleaner Fish in Salmon Aquaculture: A Review of the Evidence Base', *Aquaculture Environment Interactions*, 12 (2020), 31–44.

13   Changing Markets Foundation, 'Investing in Troubled Waters', July 2021.

14   Runar Bjørkvik Mæland, 'Study: Cleaner Fish No Magic Bullet Against Salmon Lice', Institute of Marine Research, 16 March 2020.

15   BarentsWatch, 'Salmon Lice'.

16   Liv Østevik, Marit Stormoen, Øystein Evensen et al., 'Effects of Thermal and Mechanical Delousing on Gill Health of Farmed Atlantic Salmon (*Salmo salar* L.)', *Aquaculture*, 552 (15 April 2022).

17   Mowi, 'Mowi's Scottish Sites'.

18   Cliff White, 'Norway's Salmon Industry Taking Heat for High Premature Death Rates at Sea Farms', *Seafood Source*, 10 May 2024.

19   Robert Hyde, Martin Green, Virginia Sherwin et al., 'Quantitative Analysis of Calf Mortality in Great Britain', *Journal of Dairy Science*, 103, no. 3 (March 2020), 2615–23.

20   Scottish Government/Riaghaltas na h-Alba, 'Diseases of Wild and Farmed Finfish', 11 January 2023.
21   European Union, 'Council Directive 98/58/EC of 20 July 1998 Concerning the Protection of Animals Kept for Farming Purposes', Amended 2019.
22   Just Economics, 'Dead Loss: The High Cost of Poor Farming Practices and Mortalities on Salmon Farms', February 2021.
23   Rob Edwards, 'Horror Photos of Farmed Salmon Spark Legal Threat', *Ferret*, 27 June 2018.
24   Chloe Coules, 'The Life and Death of Farmed Fish', *Animal Equality*, 5 April 2023.
25   Conservative Animal Welfare Foundation/Animal Equality UK, 'The Case for Regular Inspections and Mandatory CCTV in Fish Slaughterhouses', June 2022.
26   UK Government, 'Farm Animal Welfare Committee'.
27   Amro Hussain, 'Fish Welfare', *UK Journal of Animal Law, Action Plan for Animal Welfare Special* (August 2021), 21.
28   Helena Horton, 'No Routine Check-ups on Welfare of Fish at Slaughter, Officials Admit', *Guardian*, 23 November 2021.
29   Norwegian Seafood Council, 'Together We Are Winning the World for Norwegian Seafood'.
30   Rudresh Pandey, Frank Asche, Bård Misund et al., 'Production Growth, Company Size, and Concentration: The Case of Salmon', *Aquaculture*, 577 (2023), 739972.
31   Mowi, 'Salmon Farming Industry Handbook 2023', p.24.
32   Allison Mood and Phil Brooke, 'Estimating Global Numbers of Fishes Caught from the Wild Annually from 2000 to 2019', *Animal Welfare*, 33, e6 (February 2024), 1–19.
33   Patricia Majluf, Kathryn Matthews, Daniel Pauly et al., 'A Review of the Global Use of Fishmeal and Fish Oil and the Fish In:Fish Out Metric', *Science Advances*, 10 (16 October 2024).
34   Cows need up to 7.5 kg of food to put on 1 kg in weight. Ethan Brown, 'Feed Conversion Ratios Help Explain Meat's Outsized Climate Impact', *PBS*, 20 March 2022.
35   Changing Markets Foundation, 'Krill, Baby, Krill: The Corporations Profiting from Plundering Antarctica', 2022.
36   Stefanie Colombo, Koushik Roy, Jan Mraz et al., 'Towards Achieving Circularity and Sustainability in Feeds for Farmed Blue Foods', *Reviews in Aquaculture*, 15, no. 3 (23 November 2022), 1115–41.

37 Monika Weiss, Anja Rebelein, and Matthew Slater, 'Lupin Kernel Meal As Fishmeal Replacement in Formulated Feeds for the Whiteleg Shrimp (*Litopenaeus vannamei*)', *Aquaculture Nutrition*, 26 (23 January 2020), 752–62.

38 Giovanni Turchini, Jesse Trushenski, and Brett Glencross, 'Thoughts for the Future of Aquaculture Nutrition: Realigning Perspectives to Reflect Contemporary Issues Related to Judicious Use of Marine Resources in Aquafeeds', *North American Journal of Aquaculture*, 81 (15 September 2018), 13–39.

39 Allied Market Research, 'Aquaculture Market Size, Share, Competitive Landscape and Trend Analysis Report, by Environment, by Fish Type: Global Opportunity Analysis and Industry Forecast, 2024–2035', May 2024.

40 www.norcod.com.

41 European Commission, 'Overview of EU Aquaculture (Fish Farming)'.

42 Huaxia, 'Booming Seafood Industry Mirrors China's "Blue Granary" Development', *Xinhua*, 16 May 2024.

43 Kaimaio Lin and Jianping Wu, 'Effect of Introducing Frogs and Fish on Soil Phosphorus Availability Dynamics and their Relationship with Rice Yield in Paddy Fields', *Scientific Reports* 10, no. 21 (8 January 2020).

44 Rob Fletcher, 'Stepping Up Tilapia Welfare in China', *The Fish Site*, 11 January 2023.

14. WINDY CITY

1 National Oceanic and Atmospheric Administration, 'How Much of the Ocean Has Been Explored?'

2 Established democracies rarely go to war with each other. When they do, they usually involve a North American or Western European democracy. Fishing conflicts were reciprocated only 20 per cent of the time; other conflicts involved reciprocation half the time. Sara McLaughlin Mitchell and Brandon Prins, 'Beyond Territorial Contiguity: Issues at Stake in Democratic Militarized Interstate Disputes', *International Studies Quarterly*, 43, no. 1 (1999), 169–83.

3 Port of New Bedford, https://portofnewbedford.org/.

4 TGS Wind, 'Offshore Wind Farms in the United Kingdom'.

5 Paul Rincon, 'UK Can Be "Saudi Arabia" of Wind Power – PM', *BBC News*, 24 September 2020.

6   Block Island Wind Farm, off Rhode Island, is called by its owners Ørsted 'America's starting five'. Ørsted, 'The Starting Five', https://us.orsted.com/renewable-energy-solutions/offshore-wind/block-island-wind-farm.

7   White House, United States, 'Executive Order on Tackling the Climate Crisis at Home and Abroad', 27 January 2021.

8   White House, United States, 'Fact Sheet: Biden Administration Jumpstarts Offshore Wind Energy Projects to Create Jobs', 29 March 2021.

9   Bureau of Ocean Energy Management, 'Lease and Grant Information', April 2024.

10   Port of New Bedford, 'Economic Impact', https://portofnewbedford.org/economic-impact/.

11   Bill Eville, 'Cape Wind Pulls Out of Nantucket Sound Wind Farm Project', *Vineyard Gazette*, 2 December 2017.

12   Secretary of the Commonwealth of Massachusetts, 'The Census 2020 Low Response Score (LRS) and New Bedford, Massachusetts', 2020.

13   The European Maritime Spatial Planning Platform, 'Conflict Fiche 5: Offshore Wind and Commercial Fisheries', 27 March 2015, p.11.

14   National Oceanic and Atmospheric Administration Fisheries, '2016–2024 Humpback Whale Unusual Mortality Event Along the Atlantic Coast', 11 May 2024.

15   Will Sennott, 'Wind and the Whales', *Martha's Vineyard*, 14 June 2024.

16   There are so few right whales left – about 360 – they are all identifiable. This one was a young female known by humans as 5120. When she died, she had a rope, probably from lobster fishing, embedded in her tail. The death of a young female is particularly devastating because of the calves she could have had. Eve Zuckoff, presenter, 'The Complicated Truths About Offshore Wind and Right Whales', *Science Friday*, 29 March 2024, 17.29 minutes.

17   Oliver Millman, 'Trump Falsely Claims Wind Turbines Lead to Whale Deaths by Making Them "Batty"', *Guardian*, 26 September 2023.

The BBC fact-checked Trump's claims: Marco Silvo and Jake Horton, 'Fact-Checking Donald Trump's Claim That Wind Turbines Kill Whales', *BBC News*, 26 September 2023.

18   Alison Pearce Stevens, 'Are Offshore Wind Farms Harming Whales?' *Oceanus*, 9 May 2024.

19   Silvo and Horton, 'Fact-Checking Donald Trump's Claim'.

20 National Oceanic and Atmospheric Administration, 'NOAA Fisheries Media Teleconference on East Coast Whale Strandings', transcript, 18 January 2023.

21 National Oceanic and Atmospheric Administration, 'Fisheries and Offshore Wind Interactions: Synthesis of Science', NOAA Technical Memorandum NMFS-NE-291, March 2023.

22 Tristan Baurick, 'Trouble in the Wind: Offshore Turbine Farms Complicate Fishing, Shrimping', *Times-Picayune/New Orleans Advocate* (*NOLA*), 27 November 2021.

23 National Oceanic and Atmospheric Administration Fisheries, 'The Atlantic Sea Scallop: A Fishery Success Story', 4 February 2021.

24 Will Sennott, 'Japanese Imports Affecting New Bedford Scallop Market', *National Fisherman*, 22 March 2024.

25 Plymouth Marine Laboratory, 'UK Fishing Community Shares Its Views on Offshore Wind', 10 January 2024.

26 This chapter was written before President Trump's second presidency, and before his executive order of 20 January 2025 pausing the issuing of offshore wind leases and ordering a review of the government's leasing and permitting of existing wind projects.

15. FISH IN, FISH OUT

1 Ingrid Giskes, 'Abandoned Fishing Gear Is Killing Marine Life and Poisoning Our Oceans', *Guardian* (letters), 22 May 2022.

2 Rena Lee, president of the Intergovernmental Committee, announced that the treaty had been passed by saying, 'The ship has reached the shore'. United Nations, '"The Ship Has Reached the Shore," President Announces, as Intergovernmental Conference Concludes Historic New Maritime Biodiversity Treaty', 3 March 2023.

3 Greenpeace, 'Global Ocean Treaty: How People Power Helped Protect the Oceans', 6 March 2023.

4 High Seas Alliance, 'High Seas Treaty Ratification Tracker'.

5 Karen McVeigh, 'First WTO Deal on Fishing Subsidies Hailed as Historic Despite "Big Holes"', *Guardian*, 21 June 2022.

Select Bibliography

Allen, Irene E., and Andrew A. Todd. *Filey: A Yorkshire Fishing Town*. Self-published, 1987.

Armstrong, Chris. *A Blue New Deal: Why We Need a New Politics for the Ocean*. Yale University Press, 2022.

Balcombe, Jonathan. *What a Fish Knows: The Inner Lives of Our Underwater Cousins*. One World, 2017.

Bochel, Margaret. *Nairn Speldings*. Nairn Fishertown Museum, 1998.

Bradley, James. *Deep Water: The World in the Ocean*. Scribe, 2024.

Clover, Charles. *The End of the Line: How Overfishing Is Changing the World and What We Eat*. Ebury Press, 2004.

Clover, Charles. *Rewilding the Sea: How to Save Our Oceans*. Witness Books, 2023.

Couper, Alastair, Hance D. Smith, and Bruno Ciceri. *Fishers and Plunderers: Theft, Slavery and Violence at Sea*. Pluto Press, 2015.

Davis, Keith Granger, Glenn David Quelch, and Anik Clemens. *Eyes on the Sea: A Look into the Fisheries Observer Profession Through Stories and Creative Works*. Self-published, 2017.

des Périers, Bonaventure. *Les nouvelles récréations et joyeux devis / de feu Bonaventure Des Périers*. Guillaume Rouille, 1561.

Dickens, Charles. *On Travel*. Hesperus Press, 2009.

East Lothian Community History and Arts Trust. *The Way We Were in Musselburgh and Fisherrow*. Self-published, 1987.

Fagan, Brian. *Fishing: How the Sea Fed Civilization*. Yale University Press, 2017.

Foer, Jonathan Safran. *Eating Animals*. Penguin Books, 2010.

Franklin, H. Bruce. *The Most Important Fish in the Sea: Menhaden and America*. Island Press, 2007.

Golden, Frank, and Michael Tipton. *Essentials of Sea Survival*. Human Kinetics, 2002.

Greenberg, Paul. *The Fish on Your Plate: Why We Eat What We Eat from the Sea*. Penguin Books, 2010.

Kelly, Colum. *At Sea Awaiting Orders*. Weasel Green Press, 2020.

Kurlansky, Mark. *Salmon: A Fish, the Earth, and the History of a Common Fate*. One World, 2021.

Lancum, F. Howard. *Press Officer, Please*. Crosby Lockwood, 1946.

Lavery, Brian M. *The Headscarf Revolutionaries: Lillian Bilocca and the Hull Triple-Trawler Disaster*. Barbican Press, 2015.

Lund, Paul, and Harry Ludlam. *I Was There: When the Trawlers Went to War*. W. Foulsham, 1978.

McCormack, Fiona. *Private Oceans: The Enclosure and Marketisation of the Seas*. Pluto Press, 2017.

Murray, Donald S. *Herring Tales: How the Silver Darlings Shaped Human Taste and History*. Bloomsbury, 2015.

Nestlé, Marion. *Pet Food Politics: The Chihuahua in the Coal Mine*. University of California Press, 2008.

Nussbaum, Martha. *Justice for Animals: Our Collective Responsibility*. Simon & Schuster, 2022.

O'Hara, Glen. *The Politics of Water in Post-War Britain*. Palgrave Macmillan, 2017.

Oppian, *Halieutica*. Loeb Classical Library, 1927.

Peña-Guzman, David M. *When Animals Dream: The Hidden World of Animal Consciousness*. Princeton University Press, 2022.

Pinchin, Karen. *Kings of Their Own Ocean: Tuna and the Future of Our Oceans*. William Collins, 2023.

Priestland, Gerald. *Frying Tonight: The Saga of Fish and Chips*. Gentry Books, 1972.

Probyn, Elspeth. *Eating the Ocean*. Duke University Press, 2016.

Roberts, Callum. *Ocean of Life: How Our Seas Are Changing*. Allen Lane, 2013.

Robinson, Joe. *The Life and Times of Francie Nichol.* Allen & Unwin, 1975.

Robinson, Robb. *Trawling: The Rise and Fall of the British Trawl Fishery.* University of Exeter Press, 1998.

Subramanian, Samanth. *Following Fish: Travels Around the Indian Coast.* Atlantic Books, 2010.

Sutherland, Jon, and Diane Canwell. *Churchill's Pirates: The Royal Naval Patrol Service in World War II.* Pen & Sword, 2010.

Urbina, Ian. *The Outlaw Ocean: Crime and Survival in the Last Untamed Frontier.* Bodley Head, 2015.

Vogler, Pen. *Scoff: A History of Food and Class in Britain.* Atlantic Books, 2020.

Walton, John K. *Fish and Chips and the British Working Class, 1870–1940.* Leicester University Press, 1992.

Warner, William W. *Distant Water: The Fate of the North Atlantic Fisherman.* Penguin Books, 1984.

Wiesner-Hanks, Merry. *Mapping Gendered Routes and Spaces in the Early Modern World.* Farnham/Burlington, 2015.

Willson, Margaret. *Seawomen of Iceland: Survival on the Edge.* University of Washington Press, 2016.

Index

Aberdeen, 31, 35, 93, 95, 220, 257
African striped grunt, 4
Age of Union, 213
Ainsley, David and Jean, 180–2
albatrosses, 37
Aleutian Islands, 73, 198
almaco jack, 4
amoebic gill disease (AGD), 232–3
anadromous fish, 227
anchovies, 14, 265
angelfish, 4
Animal Equality, 234–5
antisubmarine patrols,151
Aquaculture and Fisheries Act (Scotland), 234
Archipelago, 195
Arcturus, 141–3
Armstrong, Chris, 8
artemia, 238
ASDIC, 149
Association of Professional Observers (APO), 192–3, 198
astaxanthin, 228
Aston Villa, 151
Attenborough, David, 16
Australynn, 34
automatic identification system (AIS), 109,

122, 175, 179, 200, 213, 250, 259
Balcombe, Jonathan, 130
Banks, Crista, 246, 248, 251–2
bantay dagat, 201–2
Barents Sea, 60
barotrauma, 117
barracuda, 159
barramundi, 239
Bartlett, Mike, 76
Baungartner, Mark, 247
Bay of Dakar, 5–6
BBC *Shipping Forecast*, 67
beam trawling, 121
Belhabib, Dyhia, 161
Bempton Cliffs, 30, 37, 42
Bennett, Sam Shrives, 89
Beresford, Admiral Lord, 145
Bering Sea, 65–6, 71, 198
Beverly, 76
Biden, Joe, 244
Bilocca, Lillian, 96, 98–105
Biodiversity Beyond National Jurisdiction (BBNJ), 257
BirdLife International, 36–7
Birdseye, 50-2, 206
black soldier fly larvae, 236

Blackburn Rovers, 151
Blackpool, 47
Blenkinsop, Yvonne, 98, 100
Block Island array, 245, 251
Blonk, Cor, 72
Board of Trade, 103
bonga, 160
Book of Genesis, 155
bottom trawling, 7–8, 61, 107–8, 121, 175, 197, 203
Bradford, William, 49
brains, of fish, 132
Braithwaite, Victoria, 15
Bread for the World, 168
bream, 4, 18, 227, 239
Bridlington, 32, 41, 47
Brown, Culum, 125–6, 128–30, 133
Brown, Ethan, 196
Bshary, Redouan, 128
Buarøy salmon farm, 225, 228–9, 231
Buckie, 93–4, 96
Burton, Michael, 101
buss, 26
butterfly fish, 4
by-catch, 8, 24, 61, 109, 116, 137, 189, 195, 197, 264–5, 267

Index

seabirds, 36–9
see also discards

Callaway, A. H., 150–1
Canary Islands, 156–8
Candido da Silva, Maria Ines, 165–6
Canmar *Pride*, 110
CAOPA, 155, 168
Cape Town Agreement, 72
cardiomyopathy syndrome, 233
Carlyle, Rev. Dr Alexander, 90
carp, 236, 239, 264
carrier pigeons, 147
catfish, 9, 17, 48, 62, 121, 135, 264, 267
Cetaceans, 8, 200
Chamfort, Nicolas, 87
Changing Markets Foundation, 167, 233
Chapman, J. W. B., 145–6
chitosan, 14
choke species, 117
chub, 236
Churchill, Winston, 45, 78, 150
cichlids, 16, 127, 132
clams, 129, 135, 179, 186
Clark, Eugenie, 3
cleaner fish, 127–8, 230–1
Clements, Chris, 221
coalfish (Black Jack), 48, 55
Cobban, David, 65, 71, 75, 78, 83
Cobban, Gary, 65–7, 70–1, 74–5, 83–4
cod, 6, 10, 13–14, 24, 31, 34, 60–1, 65, 98–9, 116–17, 134, 138
cod skin, 165–6
'false cod', 160
farming, 238–9
and fish and chips, 46–7, 53, 55, 58, 63
and fish fingers, 50–1
Cod Wars, 103
Cohen, Mike, 131
cold shock response, 75
Coleridge, Samuel Taylor, 17
coley, 264, 266
Colonsay, 181
Compassion in World Farming, 125
condoms, 149
Conservative Animal Welfare Foundation, 234
Cook, Alfred 'Bubba', 192–3, 199–200, 237, 258–9
copepods, 92, 238
corals, 115, 174, 178, 180, 229
Cornwall Fish Producers' Association, 118
Cornwall Seafood Guide, 121
Costello, Mark, 175
Cousteau, Jacques, 173
Cowie, Maggie, 93
crab pots, 41, 65–6
crabbers, 23, 82, 108, 121
crabs, 23–5, 28–30, 33, 71, 107, 121, 196, 237, 255, 256
Crawford, Rory, 36–8
Crisp, Thomas, 147
crocodiles, 9
Crook, Andrew, 46, 48, 63
Cullercoats, 90–1
cuttlefish, 16

Dalhousie, 146
Danny FII, 193
Darwin, Charles, 194
Davis, Keith, 192–3
de la Victoria, Jojo, 202

Deadliest Catch, 16, 71
Death at Sea, 191
Denness, Mary, 98, 100
DHA, 240
Diallo, Babacar, 164
Dickens, Charles, 110–11
Dieye, Amedi, 158
Diouf, Anta, 170–1
discards, 7–8, 36–7, 115–17, 194–5, 197, 213, 235
see also by-catch
dogfish (rock salmon), 18, 47–8, 121
dolphins, 12, 39, 115–16, 119, 197, 265
dorade, 159
dory, 26
Dulay, Santiago, 202
Duncan, Doug, 215
Dzugan, Jerry, 68–70, 73–4, 78

Earle, Sylvia, 8
Eddom, Harry, 101–3
EHA, 240
eicosapentaenoic acid, 15
electromagnetic fields, 248
emergency position-indicating radio beacons (EPIRBs), 66, 68, 70–1, 79–80
Emmy Rose, 73
Encyclopaedia of Conservation, 179
Environment Agency, 36
Environmental Defense Fund, 7
Environmental Justice Foundation (EJF), 61, 212, 214, 259
Essentials of Sea Survival, 75
Estakhri, 238
European Food Safety Authority (EFSA), 12

INDEX

European Union (EU), 116, 161, 168, 175
exclusive economic zones (EEZs), 257–8
Eyes on the Seas, 192, 195

Farm Animal Welfare Committee, 234
Farquhar, Elsie, 96
Faye, Mesa Diaba, 171
Ferdinand, Jenn, 198
Filey, 23–8, 30–1, 33, 36–8, 41–2, 259
Filey kittiwakes, 37
Financial Transparency Coalition, 10
fireworks anemones, 174
First World War, 144, 146, 148, 150
Firth of Forth, 8
fish aggregating devices (FADs), 109
fish and chips, 45–9, 54–60, 62–3, 103, 206
fish consumption, 14–15
fish farming, 12–13, 16, 19–21, 225–40
FIFO ratio, 236
fish fingers, 13–14, 17, 50–2, 59, 237
FISH Platform, 72
fish skin grafts, 165–6
fishcakes, 53, 227
Fishermen's Charter, 98
Fishers and Plunderers, 212
Fishing Life, A, 107
Fishing News, 141
'fishing pressure', 257
fishmeal, 166–9
Fishmongers' Company, 88
fishwives, 87–91, 105
Flamborough, 34

flame shells, 174, 179–80
Fletcher-Cooke, Charles, 103
flither girls, 33–4
flotation bibs, 76
flute fish, 4
Foer, Jonathan Safran, 16
French Coastguard, 142

Gallagher, Paul and Jeannette, 177
Game Cock, 146
Ganacias, Arthur, 65, 83
Garvellachs, 182
Gatt, Maria, 96
Gay, Phil, 101
George VI, King, 91
GGN numbers, 62
Ghost Fishing, 255–7
ghost fishing gear, 255–7
gill nets, 34, 37–8, 61, 114, 118, 200, 255, 257
Giskes, Ingrid, 255
glass eels, 6
glauconite, 252–3
Global Fishing Watch, 213, 259
Global Ocean Treaty, 257–8, 260
Global Seafood Alliance, 239
gobies, 132–3
Goffman, Erving, 130
Golden, Frank, 75
goldfish, 125–6, 133
'Good Fish Guide', 60, 62
Goodall, Jane, 125
Goodrick, James, 100
Graves, Cassie, 120
Greenberg, Paul, 40
Greenpeace, 163–5, 170, 173–4, 195
Gribble, Dean, 65–7, 72, 74–5, 78–80, 83–4
Grills, Evan, 83
groupers, 9, 127, 159

guanine, 166
Guardian, 107, 109–10, 114, 121–2
Gulf of Mexico, 80–1
Guo Ji 289, 213
gutting, 119

haddock, 14, 24, 31, 34, 46–8, 53, 55–6, 58, 60, 63, 227, 264–5
hake, 13, 47–8, 117, 161, 227, 263, 266
halibut, 31, 68, 134, 138, 198, 239
Hammarstedt, Peter, 213
Hansen, Eric, 248–53
Harrison, Rex, 23–8, 30–42, 108, 147, 259–60
Harvest Reaper, 114
Hawaiian longline fleet, 212
Headscarf Revolutionaries, 98–9, 103
heavy metals, 11, 265
herring, 14–15, 31, 50, 52, 88, 89, 92–6
herring girls, 92–6
Hessle Road Women's Committee, 99
Hessler, Kathy, 129
Hilbig, Reinhold, 113
HM Coastguard, 144
HMT *Arab*, 152
HMT *Fidget*, 146
HMT *Iron Duke*, 145
HMT *Rutland*, 152
HMT *Victory*, 145
Holt, Rohan, 174, 179, 183–4
hook and line, 200–1, 203
Huddersfield Town, 151
Hull, 47, 97–105, 126, 148, 150, 210

• 304

Index

Hull Fishing Vessel Owners' Association, 101
Human Rights at Sea, 189–90, 194
human trafficking, 192, 199, 212, 221
Huppert, Doreen, 111

Ibsen, Henrik, 136
Ihrsen, Fredric, 134–8
I'll Try, 147
illegal, unreported, and unregulated (IUU) fishing, 9–10, 87, 168, 199–200, 259
immersion suits, 68–71, 73–4, 83
Indian clinker, 26
Indian Ocean Tuna Commission, 199
Innes, Jean, 93
Inouye, Daniel, 212
Inshore Fishing (Scotland) Act, 178
Inter-American Tropical Tuna Commission, 195
International Convention for the Safety of Life at Sea, 72
International Council for the Exploration of the Sea, 115, 118–19
International Labour Organization, 72, 208
International Maritime Organization (IMO), 72–3, 109, 200
International Organization for Migration, 158
International Transport Workers' Federation (ITF), 208, 215, 221

International Union for Conservation of Nature (IUCN), 61, 159, 265
Inuit, 50
iodine, 60, 238, 267

Jacquet, Jennifer, 62
Japanese amberjack, 239
Java stingaree, 19
jellyfish, 183
Jenkinson, Fanny, 32
John Gray Centre, 89
Johnson, Boris, 244
Juarez, Mirna, 87

Kaierua, Eritara Aati, 189–92, 194
Kamara, Amadou, 165
Kébé, Codou, 160
kelp, 134, 138, 175
Kerecis, 165–6
Key, Brian, 132
King, Margaret H., 90, 94
Kingston Peridot, 97–8, 104
kippers, 92
Kiribati, 189–91, 194
krill, 62, 228, 236

Lady Elsa, 150
Lady Shirley, 150–1
Lake Mbane, 164
Landward, 180–1
Lavery, Brian, 99–100
Lawler, Jon, 65, 78–80, 83–4
Leicester City, 151
les jours maigres, 55
Liberian Coast Guard, 213
lice, 20, 226, 229–33, 240, 266
lifejackets, 38, 66, 68, 70, 76, 78, 80, 97, 158
see also personal flotation devices

Lifejackets for Lobstermen project, 76
Lincoln, Jennifer, 80
'line-caught', 60
ling, 31, 48
linoleic acid, 15
lobster, 13, 18, 23–5, 28–30, 33, 76, 125, 186, 237
Loch Carron, 179–80
Loch Seaforth, 232
Loch Sunart, 178–9
Loggins, Kenny, 74
longline fisheries, 7, 36, 265
Lorella, 103
lotte, 159–60
Ludlam, Harry, 148, 151
lumpfish, 229–30, 238
Lund, Paul, 148, 151
lupin kernels, 236

MacAlister, John, 186
McClure, Tom, 107–10, 113–15, 119, 121–2, 260
mackerel, 14, 16, 118, 201, 264
Madagascan lakana, 26
Madagascar, 236–7
maerl, 174
Maersk *Kendal*, 109
Magnuson-Stevens Act, 198
Mahatante, Paubert Tsimanaoraty, 236
Mainprize, Sally, 34
Malin, Joseph, 49
Mari, Francisco, 168
Marine Accident Investigation Branch, 68, 215, 219
Marine Conservation Institute, 109
Marine Conservation Society, 60, 264, 268

INDEX

Marine Ingredients Organisation, 168
marine protected areas (MPAs), 134, 138, 173–5, 178–80, 183–4, 186, 195, 200, 203, 260
Marine Protection Atlas, 134, 186
Marine Safety Education Association (AMSEA), 73–4
Marine Scotland, 178–80, 182–4
Marine Stewardship Council (MSC), 60–2, 92, 198
Mauritania, 167–8
maximum sustainable yield (MSY), 118
Mayhew, Henry, 88
Mballing, 157, 162, 170
megrim, 121
menhaden, 258
mercury, 11, 264
milkfish, 239
Miller, David, 46, 56–8, 60, 63
mines, 121, 149–50
minesweeping trawlers, 145–6, 149–50
Minkiewicz, Andrew, 245
minnows, 17
Mitchell-Rachin, Liz, 193
Monfort, Marie Christine, 87
monkfish, 118, 134, 159, 178, 246
Monterey Bay, 264–5
Montgomery, Sy, 132
moray eels, 4–5, 262
Mowi, 227–8, 232, 235
MRAG, 192–3, 195
Muir, Robert, 149
Munby, Arthur, 34
Murray, Tom, 137

Musselburgh, 91
mussels, 33, 237, 239–40, 267–8

Nagorno-Karabakh war, 165
Nairn, 91
National Federation of Fish Friers (NFFF), 45–6, 48, 52–4, 60, 63
National Federation of Fishermen's Organisations, 131
National Institute for Occupational Safety and Health, 67, 80
National Transportation Safety Board, 84
Nature Scotland, 174, 179
Ndiaye, Abdoulaye, 161–3
Nelson, 147
Nemecky, Stella, 116–17
Neptune's Army, 255
New Bedford, 78, 243–6, 249, 251
New Under Tens Fishermen's Association (NUTFA), 39–40
Newhaven, 91
Newlyn, 107, 117, 121–2
Nicholson, Thomas, 207, 219–21
Nil Desperandum, 146
Nimrod, 23, 25–6, 31, 33
No Take Zones, 181, 186, 260
Norcod, 238–9
North Pacific Observer Program, 195, 198
Northeast Center for Occupational Safety (NEC), 76
Norway Fisheries Museum, 225

Norwegian Climate and Pollution Agency, 229
Norwegian Directorate of Fisheries, 239
Norwegian Institute of Marine Research, 237, 239
Norwegian Seafood Council, 235
Nouvian, Claire, 260
nudibranchs, 5
Nunn, Fred, 255–6
Nyepi, 17

Oceana, 15, 175, 200–1, 236
octopuses, 18, 130, 132, 201
offshore wind energy, 243–53
Olivia Jean, 206–7, 209, 211, 216, 218–20
omega-3, 14–15, 60, 92, 238, 265–7
Open Seas, 173–4, 176, 178–9, 183
Oppian, 12, 18–19, 201
Orsino, 103
Orwell, George, 45
otoliths, 196
Otter Hound, 146
otters, 238
Oubre, Fr Sinclair, 80–2
Outlaw Ocean, 214
oysters, 13, 131, 174, 239, 240

Pagal, Norlan, 201, 203
Parkes, Graeme, 193
parrot fish, 4
Paul VI, Pope, 50
Pauly, Daniel, 6, 8, 10, 40, 62, 175, 186, 258
penguins, 17, 62, 236
perch, 199, 264
Percy, Jerry, 39–40

• 306

Index

personal flotation devices (PFDs), 68–70, 76, 81–2
see also lifejackets
PETA, 131
Petit, Rev. Arthur, 31
Pew Trusts, 192
Philomena, 219
pike, 129
Pipernos, Sara, 191
Pir2, 122
plaice, 31, 47, 55, 117–18, 238–9, 266
Plymouth Marine Laboratory, 251
pollock (pollack), 13, 20, 47, 51, 118, 199, 266
porpoises, 61, 115
Port Arthur Area Shrimpers Association, 80
Port State Measures Agreement (PSMA), 199
Posdaljian, Natalie, 196–7, 199
prawns, 8, 13–14, 60–1, 87, 168–9, 215, 228, 264, 267
  tiger prawns, 236, 267
proliferative gill disease, 232
Proops, Marjorie, 102–3
Public Health Act, 56
puffer fish, 4–5
Puritans, 49–50

Q-ships, 147
Quince, Joel, 219–20

Rainey, Brock, 65, 83
Rasul, Santanina, 201
rays, 5, 7, 115–16, 118, 125, 130, 160
  stingrays, 4–5, 14, 115, 132

Red Cross of Scotland, 94
Regan, John, 244–5, 247
regional fisheries management organizations (RFMOs), 20, 118, 190, 195, 198
Relano, Veronica, 175
remotely operated underwater vehicles (ROVs), 174, 176, 183–5, 226
Research, 32
Responsible Offshore Development Alliance, 248
Rigny, Graeme, 52
RMS *Britannia*, 110
Roach, Mary, 113
Roberts, Callum, 175
Robertson, Alastair, 215
Robinson, Robb, 144
Rodrigo, 103
Rose, George, 104
Ross Antares, 103
Ross Cleveland, 97, 101–2
rougets, 160
Rousseau-Gano, Seth, 65, 83
Royal National Lifeboat Institution (RNLI), 24, 70, 143–4
Royal Naval Patrol Service (RNPS), 148–52
Royal Naval Reserve (RNR) 145, 147–8, 150
Royal Navy, 111
Royal Society for the Protection of Birds (RSPB), 36–7
Ruff & Reddy, 82

sablefish, 199
Safina, Carl, 132

St Keverne, 99
St Romanus, 97–8
saithe, 134
SalmoFan, 228
salmon, 13–15, 45, 62, 169, 264, 266
  farming, 21, 225–40, 266
  fish fingers, 51
  rock salmon, *see* dogfish
salmon boats, 24–6, 28, 31, 33–4, 37–40
Saltstraumen, 134–6, 139
sardinella, 160, 162, 168, 258
sardines, 200, 265
scallop fishing
  dredging, 177–83, 185–6
  and modern slavery, 205, 209, 211–12, 216
  and offshore wind, 245–6
scallopers, 141–3
scallops, 48, 115, 174, 176–83, 185–6, 248–50
scampi, 9, 13, 61
Scandies Rose, 65–7, 70, 72–4, 79–80, 82–4
Scarborough, 24, 30–3, 41, 47, 96, 146, 185
Scarborough woof, 137
Scott, Sir Walter, 89
Scottish Creel Fisherman's Federation, 186
Scottish Salmon Company, 234
Scottish White Fish Producers Association (SWFPA), 185–6
sea bass, 227, 236, 239
Sea Beaver, 173–4, 183–4
sea cucumbers, 239
Sea Lady, 219

307 •

INDEX

sea pens, 179
Sea Search, 179
Sea Shepherd, 183, 213
sea urchins, 4–5, 39, 129, 135
Seafish, 62–3
Seafood and Fisheries Emerging Technology conference, 258
sealions, 116
seals, 39, 41, 116, 185, 236, 256
seasickness, 21, 110–13
Seaver, Barton, 49
seaweed, 39, 174, 185, 200, 229, 233, 237
Second World War, 146, 148, 243
Seiders, Pete, 77
selenium, 164, 267
Sene, Senabou, 157, 163
Senegal, 155–71, 236
sentience, 15–17, 125, 128–31
shark egg paste, 239
shark fin, 61, 190, 192
sharks, 3, 5–7, 17, 111, 115–16, 121, 126, 130, 150, 160, 197, 199
  pig horns, 226
  requiem shark, 16
  spur dogs (smooth hounds), 118
  see also dogfish
shrimp, 8, 14, 18, 80–2, 131, 227, 237, 255
Silver Cod Trophy, 99
Singer, Peter, 131
skate, 31, 47, 118, 181, 197
  flapper skate, 184–5
'Slavery at Sea' documentary, 222

small pelagics, 160–2, 236
Sneddon, Lynne, 132
sole, 31, 239
Sorensen, Julie, 76–7
sprag, 47
Staaterman, Erica, 247
starfish, 4, 115–16, 118, 205
Statistical Account of Scotland, 90
Steele, Joe, 150
Stella Maris, 80, 206, 210, 215–17, 262
Stinson, Davy, 178, 180
Stoakes, Charles George, 150
Storeblå acquaculture centre, 225, 231
Strachan, Raymond, 142–3
Strange, William, 131
Strathmore, 215
Stump, Chris, 196
subsidies, 259–60
Sunniside Local History Society, 91
Sunshine, 146
superstitions, 30, 69, 99
surströmming, 92
sushi, 6, 12, 15
Sustainable Food Trust, 233
swim bladders, 116–17, 166
'swinging the lantern', 41

Tañon Strait, 200–2
Tarantula, 146
Tasawu Kayar Collective, 164–5, 170
Task Force on Human Trafficking in Fishing in International Waters, 212
Tate, Harry, 148
The Gambia, 167–8

thieboudienne, 159–60
thiof, 159–61
Thompson, Paul, 96
thornback, 31
tilapia, 236–7, 240, 266–7
  tilapia skin, 166
Tipton, Michael, 75
TN Trawlers, 207, 209, 218–21
Tory, Sarah, 192
total allowable catch (TAC), 118
Touba Protéine Marine (TPM), 174
Trent, Steve, 61, 214, 259
trimethylaminuria, 120
trout, 132, 233, 239, 260, 264, 266
  sea trout, 24, 29, 31, 34–5
Trump, Donald, 247
tuna, 6–7, 11–14, 16–18, 109, 119–20, 161, 200–1, 212, 236–7, 239, 258, 264–5
  bluefin, 11–13, 17, 226, 260–1
  farmed, 12–13
  and fisheries observers, 189–90, 192, 195
  skipjack, 11, 265
turbot, 31, 117
Turtle, Joan, 117
turtles, 4–5, 7–8, 115–17, 197, 255

U-boats, 146–7, 150–1
UK Animal Welfare (Sentience) Act, 15
Ullswater, 150
UN Food and Agriculture Organization (FAO), 7–9, 72–3, 118, 162, 167, 205, 228, 239, 265
UN Ocean Conference, 258

Index

UNAPAS, 161
UNCTAD, 7
Underdown, Nick, 173
Unique Vessel Identifier, 259
Urbina, Ian, 214
US Bureau of Ocean Energy Management, 244, 247, 251
US Coast Guard, 66, 69–70, 72–4, 76, 81–4, 193–4
US Department of Labor, 213
US Fisheries Conservation and Management Act, 194
US Food and Drug Administration (FDA), 11, 264
US National Oceanic and Atmospheric Administration (NOAA), 7, 195, 197–8, 247–9, 251

Utopia 56, 142
venting, 117
vessel monitoring system (VMS), 181–2, 259
Victoria 168, 192
Viem, Borghild, 134–9
Vineyard Wind, 244–8, 250–2
vitamin D, 92, 267

Walker, Tony, 227, 240
Walton, John K., 47
Warner, William, 122
Waters, Hall, 107
Watson, Angus, 52
Welfare of Animals at the Time of Killing Regulations, 234
Western Central and Pacific Fisheries Commission (WCPFC), 190–1
whales, 17, 115–16, 121, 174, 197, 245, 247

fossilized whale poo, 252–3
whiting, 265
Wilberforce, William, 105
Wilson, Harold, 100
Win Far No 636, 190–1, 194
Winner, Joseph, 152
wolffish, 133–9, 239
Women in Ocean Science, 197
Women in the Seafood Industry, 87
World Trade Organization, 17, 259–60
Worm, Boris, 175
wrasse, 126–9, 229–30, 238
WWF, 8, 17, 116, 192, 199, 237, 255, 258

zebrafish, 129
zinc, 60, 267